NORTH CAROLINA
STATE BOARD OF COMM
LIBRARI
SAMPSON TECHNICAL COLLEGE

S0-BHX-633

TK
7879.2
R573
1982

No. 1390
$14.95

10-28-82

186663

THE
VOM–VTVM HANDBOOK
BY JOSEPH A. RISSE

TAB **TAB BOOKS Inc.**
BLUE RIDGE SUMMIT, PA. 17214

FIRST EDITION

FIRST PRINTING —1982

Printed in the United States of America

Reproduction or publication of the content in any manner, without express permission of the publisher, is prohibited. No liability is assumed with respect to the use of the information herein.

Copyright © 1975, 1969, and 1963 by TAB BOOKS Inc.

Library of Congress Cataloging in Publication Data

Risse, Joseph A.
      The VOM-VTVM handbook.

      Includes index.
        1. Voltohmmeter.   2. Vacuum-tube voltmeter.
I. Title.   II. Title: V.O.M.-V.T.V.M. handbook.
TK7879.2.R573   621.37′42   81-18410
ISBN 0-8306-0079-5   AACR2
ISBN 0-8306-1390-0 (pbk.)

Preface

In the past, the volt-ohm-milliammeter (vom) and the vacuum-tube voltmeter (vtvm) have been the test instruments used most widely in electrical and electronics work. The vtvm differs from the vom in that the vtvm includes vacuum-tube circuits for amplification of low-level signals and for otherwise adding to the versatility of the instrument. In recent years, other forms of electronic voltmeters have appeared and have become very widely used. These more-recent instruments employ solid-state devices such as transistors or ICs (integrated-circuits) rather than vacuum tubes, and they are generally battery-operated (or offer the choice of either battery or power-line operation); thus they are advantageous to use when portability is required. The electronic instruments that employ solid-state devices (transistors) are known by such names as "electronic voltmeters" (evm's), "solid-state voltmeters" (ssvm's), "field-effect-transistor voltmeters" (FET vm's), "field-effect-transistor volt-ohm-milliammeters" (FET vom's), and so on.

The instrument mentioned so far are classified as *analog* instruments; they employ a pointer and a calibrated scale. The pointer indicates the value being measured along the scale which is marked off from zero to some maximum value. Another general type of instrument which must be considered of equal importance with the analog type is the *digital* instrument. These are referred to as digital voltmeters (dvm's) or digital multimeters (dmm's).

Because of the widespread use of analog and digital instruments, it is important to have a thorough knowledge of their features, use, operation, and maintenance. The main purpose of this book is to provide as much of this knowledge as possible. Discussions of the circuits and of the uses, care, and repair of the instruments are discussed. The major features of the various types are covered, and the advantages and disadvantages are described.

The coverage is practical and relatively basic; the occasional arithmetic employed in the discussions involves only simple calculations. For the benefit of students, hobbyists, experimenters, and beginners, the beginning of the book reviews basic concepts necessary for understanding electrical measurements. In addition, for the benefit of students and for technicians preparing for employment, licensing, or certification tests, review questions follow each chapter, and answers are at the end of the book.

The author wishes to express his thanks to the International Correspondence Schools division of INTEXT for the use of their library and laboratory facilities; to Pursell Electronics for providing the opportunity to evaluate several instruments from various manufacturers; to Dr. Eugene A. McGinnis, Chairman, Department of Physics, University of Scranton, for advice and guidance; to the manufacturers who freely provided technical research information related to their particular products; to Bill Risse for photographic assistance; and to Ron Lettieri, Tobyhanna Army Depot, for valuable comments.

Dedicated to my wife
Anne Stegner Risse
and to our children
Bill, Sally, Joe, Jane, and Ed

Contents

CHAPTER 1

CHAPTER 2

CHAPTER 3

CHAPTER 4

CHAPTER 5

Uses of VOMs
and VTVMs

The most widely used test instrument in electrical and electronics work is the volt-ohm-milliammeter (vom). This instrument is used by electricians, technicians, experimenters, teachers, scientists, engineers, inventors, and students. Vom's find wide application in troubleshooting; electrical maintenance; design; production work; radio, tv, and appliance servicing; and in many other areas. Electronic vom's or multimeters, including the solid-state, vacuum-tube, integrated-circuit, and digital types, are also widely used. The student or technician usually starts out using a nonelectronic vom but soon adds one or more electronic types to his personal array of instruments.

Since vom's and the various emm's (electronic multimeters) are so widely used, it is important for the electronics worker, whether he be student, technician, teacher, engineer, or experimenter, to understand their use, care, maintenance, and principles of operation.

The topics discussed in this book cover the types of vom's and emm's most widely used. Detailed discussion of a particular instrument does not necessarily imply that it is the best one or the only one. In cases where an instrument described is not the latest model, a later version will generally be the same except for minor improvements.

However, it is not the main purpose of this book to discuss particular instruments, but rather to explain the major features common to these instruments. It will tell you what features should be considered before purchasing a new instrument or tell you how you can best use the one you have.

REVIEW OF ELEMENTARY ELECTRICITY

At this point, before beginning with detailed dicsussions about vom's, it might be worthwhile to review some of the principles of electricity—especially those principles that are important in understanding vom's.

You are probably well aware that all matter is composed of molecules, which are made up of atoms. The atoms consist of neutrons, protons, and electrons. The theory that electricity is the flow of electrons through a conductor (such as a wire) is now pretty well established. *Electrons* are negative charges of electricity. They are repelled by other negative charges of electricity, and attracted toward positive charges.

Current

A movement of electrons constitutes an electrical *current*. The number of electrons that move during a given period of time determines just how much current is flowing. Current is measured in amperes. Currents less than 1 ampere are generally measured in milliamperes or microamperes. The conversion between amperes (A), and milliamperes (mA), and microamperes (μA) is rather simple; 1 ampere equals 1000 milliamperes or 1,000,000 microamperes. To convert from:

Amperes to milliamperes, multiply amperes $\times$ 1000.
Amperes to microamperes, multiply amperes $\times$ 1,000,000.
Milliamperes to amperes, divide milliamperes by 1000.
Milliamperes to microamperes, multiply milliamperes $\times$ 1000.
Microamperes to amperes, divide microamperes by 1,000,000.
Microamperes to milliamperes, divide microamperes by 1000.

Examples:

Converting amperes to milliamperes:
   0.0001 ampere = 0.0001 $\times$ 1000 = 0.1 milliampere
Converting amperes to microamperes:
   0.0001 ampere = 0.0001 $\times$ 1,000,000 = 100 microamperes
Converting milliamperes to amperes:
   0.1 milliampere $\div$ 1000 = 0.0001 ampere

In practical situations, the units are seldom converted. Instead, the most convenient unit for the amount of current is used. For example, the use of 10 amperes is preferred to the use of 10,000 milliamperes or 10,000,000 microamperes. Similarly, the use of 5 milliamperes is preferred to the use of 0.005 ampere.

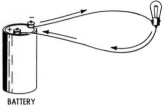

Fig. 1-1. Direct-current circuit.

BATTERY

DC and AC Voltage

Voltage is the energy that causes electrons to move. Usually the source of voltage is a battery or an electric generator. There are two basic kinds of voltage; one is *dc voltage,* and the other is *ac voltage.*

Direct-current (dc) voltage causes the electrons to move from the negative side of the voltage source, through one of the two wires required, through the circuit being powered, to the second wire, and back to the positive side of the voltage source. An example of such a dc circuit, a battery providing the power for a lamp bulb, is shown in Fig. 1-1. The movement of electrons is continuous from negative to positive, until the voltage is removed by opening the circuit.

Alternating-current (ac) voltage also causes electrons to move from negative to positive. The difference between ac and dc voltage, however, is that the ac polarity keeps changing at a regular rate, as does the amplitude (magnitude) of the voltage, while the dc polarity remains constant. At first, one of the two terminals of the ac voltage source may be negative and the other positive, as shown for terminals A and B in Fig. 1-2A; then the one that was negative becomes positive, and the one that was positive becomes negative, as in Fig. 1-2B. Side A starts from zero voltage (point 1 in Fig. 1-2C), builds up to a maximum positive value (point 2), decreases to zero again (point 3), builds up to the maximum negative value (point 4), falls back to zero (point 5), and rises to maximum positive value (point 6). This sequence is then repeated. The result is that the electrons flow through the wire first in one direction, then stop, flow in the other direction, reverse again, and so on until the circuit is opened.

The number of times per second that a cycle of these variations (zero to positive to zero to negative to zero) occurs is known as the *frequency* of the ac voltage. Frequency is sometimes specified in *cycles per second* (*cps*) or simply *cycles.* But the term *hertz* (abbreviated Hz) has become the preferred unit for frequency. Therefore, in some cases you may encounter frequency specified in terms of cycles per second; in others it may be specified in terms of hertz. One hertz is equal to one cycle per second. The value of ac voltage

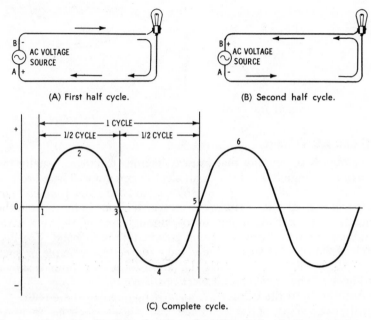

(A) First half cycle.

(B) Second half cycle.

(C) Complete cycle.

Fig. 1-2. Alternating directions of electron flow through the circuit connected to ac generator for voltage source.

supplied by the utility company to the average home is about 120 volts, and its frequency is usually 60 Hz.

Resistance

The opposition to the flow of electrons through a material is known as *resistance*. Resistance is low through a wire and through most metals and other conductors. Through nonconductors, or *insulators*, resistance is high. Resistance is measured in *ohms*.

Some components in electrical circuits consist of materials designed to have a specific value of resistance; these components, known as *resistors*, are available in many sizes and values of resistance. The amount of current a resistor must carry usually determines its physical size. Conductors, such as wires, generally have a low resistance—a typical figure would be 1 to 10 ohms for a 100-foot piece of wire.

Relationship of Voltage, Current, and Resistance

In a given electrical circuit, the higher the applied voltage, the greater the amount of current. The lower the voltage, the less the current. Similarly, the greater the resistance, the lower the current, and vice versa.

The relationship of current, voltage, and resistance is given by Ohm's law as:

$$E = IR$$

where,

E is the value of the applied voltage, in volts,
I is the amount of current, in amperes, flowing in the circuit,
R is the value of the resistance, in ohms.

The preceding equation states that the voltage is equal to the current multiplied by the resistance. Thus, if the values of the resistance and the current are known, then their product gives the value of the applied voltage.

If any two of the three quantities are known, the third can be determined. For instance, if the voltage and current are known, the equation for determining the resistance is:

$$R = \frac{E}{I}$$

Or, resistance is equal to voltage divided by current.

And, if the resistance and the voltage are known, the equation for the current is:

$$I = \frac{E}{R}$$

Or, current is equal to voltage divided by resistance.

Power

The amount of power absorbed in an electrical circuit is usually stated in watts. The power in watts can be found from the equation:

$$P = I^2R$$

where,

P is the power in watts.

Or, the power equals the value of the current squared times the resistance.

The power can also be found if the current and voltage are known:

$$P = EI$$

Or, if the voltage and the resistance are known:

$$P = \frac{E^2}{R}$$

In many cases it is important to use an electrical component that is large enough in physical size to dissipate the heat due to the flow of electrical current through the component. Therefore, resistors have a rating in watts. Resistors are commercially available in a variety of wattages. In designing circuits, usually a resistor having twice the wattage rating needed is selected. This is to allow for the possibility that the resistor will operate at a higher temperature than calculations would indicate, due to its location beneath a chassis or inside a cabinet, where the heat is not readily carried away by convection, conduction, or radiation.

Suppose that in a circuit with a voltage source of 100 volts, 0.005 ampere of current is required. The value of the desired resistor is determined from the equation given earlier:

$$R = \frac{E}{I} = \frac{100}{0.005} = 20,000 \text{ ohms}$$

Next, to determine its rating in watts, use the formula:

$$P = EI = 100 \times 0.005 = 0.5 \text{ watt}$$

Using either formula $P = EI$ or formula $P = I^2R$ would have given the same answer.

In an open space, a ½-watt resistor could be used; but if the circuit is enclosed or if the escape of heat is restricted in any manner, a 1-watt resistor would normally be used.

Characteristics of Alternating Current

The electrical principles and laws just given for dc apply equally to ac. That is, 100 volts ac will cause 0.005 ampere to flow through 20,000 ohms, the same as 100 volts dc will.

It is important, however, to realize the special characteristics of ac. Since ac ranges from zero to maximum positive to maximum negative during each cycle, its average value for each cycle is zero. Of course, as far as effects on an electrical circuit are concerned, this average value of zero does not mean that ac has no effect. When an ac voltage forces electrons through a circuit first in one direction and then in the other, it creates heat or causes other effects, as a dc voltage does. However, the *effective* value of alternating current is not actually its *maximum* value. The effective value of ac is related to its maximum value by the factor 0.707, as indicated in the illustration of Fig. 1-3.

$$E \text{ (ac effective)} = E \text{ (ac maximum)} \times 0.707$$
$$I \text{ (ac effective)} = I \text{ (ac maximum)} \times 0.707$$

In the other direction, the maximum value is related to the effective value by the factor 1.414.

$$\text{Maximum} = \text{Effective} \times 1.414$$

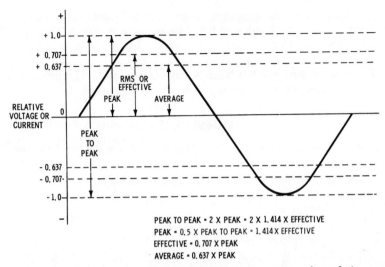

Fig. 1-3. Relationship among peak-to-peak, effective, and average values of sine-wave alternating voltage or current.

Thus, as an example, the effective value of an ac voltage having a value of 150 volts for its positive and negative peaks can be calculated as:

$$150 \times 0.707 = 106 \text{ volts}$$

An ac voltage having an effective value of 110 volts has a maximum value of:

$$110 \times 1.414 = 155.5 \text{ volts}$$

Another term frequently used for maximum values is *peak* value, and another term commonly used for effective value is *rms* (root-mean-square) value.

The value of an ac voltage or current is sometimes given in terms of its *peak-to-peak* value. It is obvious that the peak-to-peak value of a symmetrical ac wave is twice its peak value:

$$\begin{aligned} \text{Peak-to-peak value} &= 2 \times \text{peak value} \\ &= 2 \times 1.414 \times \text{effective value} \\ &= 2.828 \times \text{effective value} \end{aligned}$$

Although the average value for a complete ac cycle is zero, each half cycle of an ac wave has a specific *average* value. The average value for a half cycle of ac is given by:

$$\text{Average value} = 0.637 \times \text{peak value}$$

13

For many practical purposes, ac voltage or current is referred to as having an average value of 0.637 times its peak value. This is not necessarily restricted to applying to only a half cycle at a time; but for other purposes it is important to keep in mind that the average value of an ac sine-wave voltage or current is zero. This will become apparent later when meter movements are discussed.

QUESTIONS

1. If a sine-wave voltage has a peak value of 150 volts, what is the average value for a half cycle?
2. What is the average value of a full cycle of ac voltage?
3. What is the effective, or rms, value of an ac voltage that has a peak value of 440 volts?
4. What are electrons? Briefly describe their action in an electric circuit.
5. What constitutes an electric current?
6. Describe the relationship between the following: amperes and milliamperes; microamperes and amperes.
7. Convert 50 milliamperes to amperes; convert 0.00005 ampere to microamperes.
8. What causes the movement of electrons in an electric circuit?
9. Name two sources of voltage.
10. Describe the direction of movement of electrons in a dc circuit supplied by a battery.
11. Describe the movement of electrons in an ac circuit.
12. What term other than "cycles per second" is used in designating frequency?
13. What is the usual power-line frequency supplied by electric utilities?
14. What name is given to the opposition to the flow of electrons in a material, and in what units is it measured?
15. What is Ohm's law?
16. What form of Ohm's law is used if the unknown quantity is (a) current, or (b) resistance?
17. Determine the resistance of a circuit in which 20 volts causes a current of 80 milliamperes.
18. How much current will there be in a circuit having 25 ohms of resistance if 100 volts is applied to the circuit?
19. How much voltage is required to cause 3 amperes to flow through 60 ohms?
20. Calculate the power in a circuit in which 50 volts is applied and 4 amperes is flowing.
21. Determine the peak ac voltage if the rms value is 150 volts.
22. If the peak value of a sine wave is 125 volts, what is the peak-to-peak value?

CHAPTER 2

VOMs

The designers and manufacturers of vom's try to make them as versatile and convenient to use as possible. The vom is a single-package instrument with which voltage, current, and resistance can be measured. The ease with which the instrument can be used is usually very important. Also important are the accuracy of the instrument and the ranges of measurement of which it is capable.

WHAT A VOM CAN DO

The basic design of the vom is shown in Fig. 2-1. The meter is the major part of the vom. The pointer of the meter indicates the value of the voltage, current, or resistance being measured. The voltage-

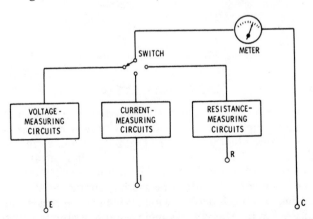

Fig. 2-1. Basic plan of a vom.

measuring section provides for both ac and dc voltage measurement over any one of several ranges. The resistance-measuring circuit has several ranges of measurement. In Fig. 2-1, terminal C is common for all measurements; the unknown voltage is connected between E and C, the unknown current between I and C, and the unknown resistance between R and C.

Generally, a vom can measure voltages up to 5000 or 6000 volts ac or dc, up to 5, 10, or 15 amperes of current, and up to 1 to 10 megohms of resistance. With some vom's it is also possible to measure decibels and power.

The basic components of the vom, as shown in Fig. 2-2, are a meter; test leads; a function switch; and voltage-, current-, and resistance-measuring circuits.

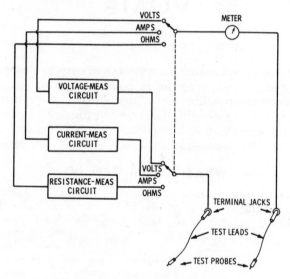

Fig. 2-2. Basic components of the vom.

BASIC VOM SYSTEM

In some vom's the switch is not included—in this case, it is necessary to move one of the test leads each time a different function is to be performed by the meter. In Fig. 2-2, if there were no switch, the right-hand test lead would be left where it is at all times and the left-hand lead would be plugged into a jack connected to the resistance-, current-, or voltage-measuring circuit, as desired. Most vom's do include the function switch, which is the reason for showing it in the figure. Many vom's include both a function switch and a range switch. The range switch is used for selecting that scale on the vom that will provide the most convenient-to-read indication for the value of current, voltage, or resistance being measured. An example of a vom having both a function switch and a range switch is shown in Fig. 2-3.

Courtesy Simpson Electric Company

Fig. 2-3. Simpson Model 260 vom.

Meter of the VOM

To understand the vom, it is necessary to understand its components. The first one we will consider is the meter, or meter movement. The meter includes a scale and a pointer, or indicator, both attached to a mechanism called a movement. When the vom is not being used, the pointer remains at the 0 side of the calibrated scale of the meter as in Fig. 2-3. When current flows through the movement, the pointer moves to some position and remains there, if the flow of current is steady, until the current changes in value. Just how much the pointer is moved, or deflected, depends on how much current is flowing through the movement.

It should be kept in mind that whether voltage, current, or resistance is being measured, it is always current that deflects the pointer.

Principle of the Meter Movement

The type of meter movement used in most vom's is known as the d'Arsonval movement. This movement includes a permanent magnet and a coil to which a pointer is attached. The pointer rotates in the field of the magnet when current passes through the turns of the coil. The principle of the d'Arsonval movement is shown in Fig. 2-4. The coil is wound about a soft-iron core, or armature. The current to be measured flows through the coil, setting up a magnetic field around the coil, which opposes the field of the magnet. The coil is rotated clockwise by these repelling fields, pulling against a spring which, when no current is passing through the coil, holds the pointer at zero on the calibrated scale. The greater the current, the greater the force turning the coil.

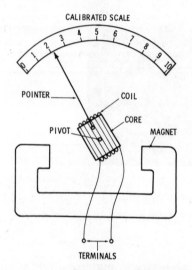

Fig. 2-4. Principle of the d'Arsonval meter movement.

The basic construction of the d'Arsonval movement is shown more clearly in Fig. 2-5. Pole pieces on the permanent magnet decrease the magnetic gap between the magnet and the coil core. Spiral springs hold the pointer, which is fastened to the coil, at zero when no current is flowing. The core of the coil does not rotate; it is stationary in order to keep the pointer assembly as lightweight as possible. The pivot usually rides in jewel bearings to reduce friction.

Since about 1960, many d'Arsonval movements employ taut-band suspension. This is a design feature that affords greater sensitivity in a meter movement; that is, it allows full-scale deflection with a lower value of current. In taut-band suspension, shown in Fig. 2-6, the pointer is suspended by a band which is a short, very thin, narrow strip of special alloy tightly suspended on special spring terminals. The torsion action of taut-band suspension makes the spiral spring unnecessary.

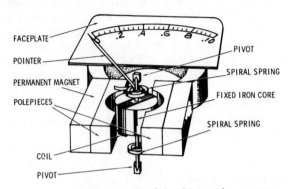

FACEPLATE
POINTER
PERMANENT MAGNET
POLEPIECES
COIL
PIVOT
PIVOT
SPIRAL SPRING
FIXED IRON CORE
SPIRAL SPRING

Fig. 2-5. Construction details of the d'Arsonval movement.

When current flows through the meter movement, the suspension band is twisted in proportion to the amount of current. When the current again becomes zero or is decreased, the pointer moves to zero or to the left. The taut-band suspension feature provides the advantage of greater sensitivity, as already mentioned, and also greater resistance to damage from both electrical overload and mechanical shock. It also responds more reliably to different frequencies. The taut-band suspension has no bearings, jewels, or spiral springs; thus, there is extremely little friction which makes possible its great sensitivity.

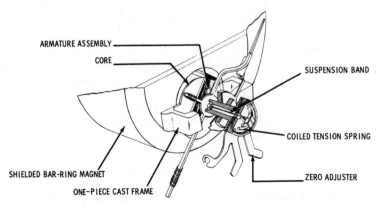

ARMATURE ASSEMBLY
CORE
SUSPENSION BAND
COILED TENSION SPRING
SHIELDED BAR-RING MAGNET
ONE-PIECE CAST FRAME
ZERO ADJUSTER

Courtesy Triplett Electrical Instrument Co.

Fig. 2-6. Principle of tant-band suspension movement.

Meter-Movement Precautions

If a meter movement becomes defective, it is usually wise to return it to the manufacturer for repair, or even to replace the movement completely without attempting repair. It is seldom possible for anyone other than a specialist to repair a meter movement properly. Many technicians have learned this the hard way. The typical meter movement is a precision assembly and includes the motor portion which is comprised of the coil, core, jewels, spiral springs, and pointer. The zero-adjuster arm, which is usually accessible from outside the meter case, is the only adjustment that the user should make. It is usually set by turning an eccentric screw on the outside of the meter case so that the pointer indicates zero when no current is flowing through the meter.

Meters are designed for various values of full-scale sensitivity. The full-scale sensitivity of a meter means the value of current that will cause the meter to deflect to its maximum indicated value. For example, 50 microamperes (0.00005 ampere) is the value that will deflect the pointer of a moderately sensitive meter to full scale. If this 50-microampere current flows through the meter when a 250-millivolt voltage is applied to the meter terminals, we can calculate the resistance (R) of the meter movement as follows:

$$R = \frac{E}{I} = \frac{0.25}{0.00005} = 5000 \text{ ohms}$$

The less current required to deflect a meter full scale, the more sensitive the meter is said to be. Some meter movements will deflect full scale with much less than 50 microamperes; many require a higher current.

CURRENT-MEASURING CIRCUIT

It is possible to measure current in a circuit by using only a meter movement if the amount of current flowing does not exceed the rated current of the meter movement. To illustrate, a 50-microampere meter in the circuit of Fig. 2-7A will be deflected to half scale. Thirty volts applied to 1,200,000 ohms (the 1,195,000-ohm resistor and the 5000 ohms of the meter movement) causes 25 microamperes to flow through the meter. This is expressed as:

$$I = \frac{E}{R} = \frac{30}{1,200,000} = 25 \text{ microamperes}$$

If the battery should be reduced to 15 volts, the meter will be deflected to ¼ of full scale; if it is reduced to 7.5 volts, the meter will deflect to ⅛ of its full-scale reading. This process can be con-

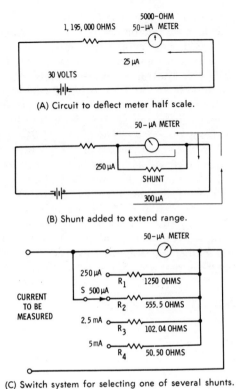

(A) Circuit to deflect meter half scale.

(B) Shunt added to extend range.

(C) Switch system for selecting one of several shunts.

Fig. 2-7. Method of using meter for current measurements.

tinued until the voltage applied will become so low that the movement of the pointer will no longer be discernible. The lowest voltage at which the movement of the meter pointer is significant will be the lower limit of usefulness. The upper limit of usefulness of the meter is, of course, the point where the battery voltage exceeds 60 volts, since beyond this voltage level the current flowing in the circuit will exceed the maximum rating of the meter. Permitting too much current to flow through a meter can cause the coil to burn out or can otherwise damage the meter.

The meter in the circuit of Fig. 2-7A can be used to measure currents greater than 50 microamperes. To do this, a *shunt*, or parallel path, is provided around the meter to carry a known portion of the current so that the current through the meter is 50 microamperes or less. This shunt is a resistor, as shown in Fig. 2-7B, with a value that must be accurately selected. For example, if a shunt of 1000 ohms is placed across the meter (Fig. 2-7B), five times as much current will pass through the shunt as will pass through the meter, since the resistance of the meter is 5000 ohms. Thus, if the meter is reading full scale, a 50-microampere current is going through the meter; 50×5, or 250, microamperes is going through the shunt; and the total current is 300 microamperes. Similarly, if the shunt has a value that is $\frac{1}{50}$ of the meter value, or 100 ohms, the current through the shunt will be 50×50, or 2500 microamperes at full scale, and the total circuit current will be 2550 microamperes.

In an actual vom, any one of several shunts can be switched into the current-measuring circuit, as shown in Fig. 2-7C. This is basically the scheme used in the vom that you might be using now or in the future. The selection of values of 1250 ohms for shunt R1, 555.5 ohms for R2, 102.04 ohms for R3, and 50.5 ohms for R4, provides additional ranges of 0 to 250 microamperes, 0 to 500 microamperes, 0 to 2.5 milliamperes, and 0 to 5.0 milliamperes, respectively. In many practical vom's, the shunt-switching arrangement makes it possible to measure dc currents as high as 10, 12, or 15 amperes or more. For these higher ranges, the shunts are pieces of wire or strap with resistances that are accurately selected.

DC VOLTAGE-MEASURING CIRCUIT

If 250 millivolts is applied to the terminals of a 50-microampere meter movement having a resistance of 5000 ohms, the meter will deflect full scale. The meter could be used by itself to measure voltages up to 250 millivolts. If a voltage is applied to the meter terminals and if the pointer deflects to its halfway point, it can be assumed that the voltage has a value of 125 millivolts.

Of course, with a typical vom it is possible to measure voltages much greater than 250 millivolts. Extending the voltage-measuring capability of a meter movement is made possible by adding a multiplier resistor in series with the meter. In Fig. 2-8A, the multiplier resistor shown has a value of 15,000 ohms. When this is added to the 5000 ohms of the meter, it totals 20,000 ohms. Since the meter requires 50 microamperes for full-scale deflection, it can easily be determined that the voltage required to provide this full-scale deflection is:

$$E = I \times R = 0.00005 \times 20,000 = 1 \text{ volt}$$

Thus, with the 15,000-ohm multiplier, a meter circuit capable of measuring up to 1 volt full scale is obtained. To add a range for measuring up to 10 volts, the total resistance of the circuit must be increased 10 times, or increased to 200,000 ohms. The value of the multiplier, then, would be $200,000 - 5000$, or 195,000 ohms. Similarly, for a 100-volt full-scale range, the multiplier must have a value of $2,000,000 - 5000$, or 1,995,000 ohms. A switch is normally used, as in Fig. 2-8B, for selecting the desired full-scale voltage ranges of 1 volt, 10 volts, or 100 volts.

In a practical vom, series multiplier resistors are used to make it possible to measure voltages up to 5000 or 6000 dc volts or more. Measurement of voltages even higher than this is possible, but usually this is not provided in a vom because of the danger of breakdown of components, arc-over between components and wiring, and danger to the user. The range of a vom can be increased to measure higher voltages by use of a high-voltage probe, as will be explained later in this book.

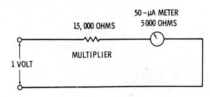

(A) Method to increase voltage range of meter.

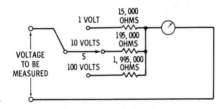

(B) Switch permits selection of desired voltage-range multiplier resistor.

Fig. 2-8. Basic vom voltage-measuring circuit.

CIRCUIT FOR MEASURING AC VOLTAGE

In a vom, the basic circuit generally used for measuring ac voltage is essentially the same as that used for measuring dc voltage. The main difference is that for ac voltage measurement, a rectifier is included in the circuit.

The rectifier may be either a half-wave rectifier, as in Fig. 2-9A, or a full-wave rectifier. Often a bridge circuit, as in Fig. 2-9B, or a double half-wave rectifier, as in Fig. 2-9C, is used. Usually, but not always, a copper-oxide rectifier is employed, rather than the selenium, silicon, germanium, or vacuum-tube types.

A rectifier permits current flow almost entirely in one direction. Therefore, it changes the ac voltage to be measured to a dc voltage in the form of a series of half sine waves. Thus, current flows through the d'Arsonval meter in one direction only, just as it does for dc measurement.

Because of the additional resistance of the rectifier in the circuit and because the current is not continuous for ac measurement, it is usually necessary for the designer of a vom to use multipliers in the ac voltage-measuring circuit that are different from those used for dc voltage measurement. This would not be necessary if different calibrated scales on the faceplate of the meter were used for dc and ac. However, it might be confusing to search through the maze of calibrations on the faceplate for the desired ac or dc scale. Thus, designers generally agree that it is preferable to design the meter so that the same voltage scales can be used for both ac and dc.

The half-wave rectifier circuit shown in Fig. 2-9A is actually seldom employed in a high-quality vom. The only important reason for this is that copper-oxide and other solid-state rectifiers conduct some current on the negative half cycles. Assume that a certain copper-oxide rectifier has a resistance of 200 ohms in one direction. This is the direction in which it would best conduct on the positive

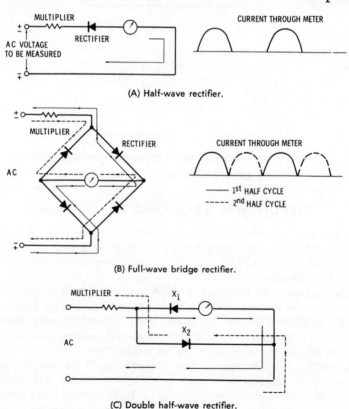

(A) Half-wave rectifier.

(B) Full-wave bridge rectifier.

(C) Double half-wave rectifier.

Fig. 2-9. Basic rectifier circuits in measurement of ac voltage.

half cycles. Then, typically, it would have a resistance of 100,000 ohms in the other direction. Therefore, on negative half cycles it would conduct very little current in most circuits. But in a vom circuit, the multiplier resistors have fairly high values on all but the lowest voltage ranges, and the reverse current may be appreciable compared to the forward current.

To illustrate, in Fig. 2-9A the resistance of the multiplier, rectifier, and meter in the forward direction might be 2,000,200 ohms on the 100-volt range, and in the reverse direction the resistance might be 2,100,000 ohms. Then, on positive half cycles the current would be very close to 50 microamperes. But on negative half-cycles, instead of the current being practically zero, it would be 100 divided by 2,100,000, or approximately 48 microamperes. The 50 microamperes going through the meter for all the positive half cycles would tend to deflect the pointer to full scale, and the 48 microamperes going the opposite direction through the meter on negative half cycles would tend to swing the pointer nearly as much in the other direction. The two opposing forces would be occurring only $\frac{1}{120}$ of a second apart, so the net effect on the meter reading would be the difference between the forward and reverse current, or $50 - 48 = 2$ microamperes. Rather than causing the meter to indicate 100 volts, then, this 2 microamperes would cause it to indicate only 4 volts on the 100-volt scale.

Whenever half-wave rectifier circuits are employed in vom's, they are often the double half-wave type shown in Fig. 2-9C. For this circuit, the overall measuring circuit has approximately the same resistance for both the half cycles of measured ac; but on the reverse, or negative, half cycles, the second rectifier (X_2) shunts the reverse current around the meter, thus preventing any appreciable part of the forward, or positive, half cycles from being cancelled.

BASIC VOM CIRCUITS FOR RESISTANCE MEASUREMENT

The basic circuits used in vom's for measuring dc current and dc and ac voltage have been discussed. The remaining major function of a vom is to measure resistance. Since resistance is measured in ohms, a resistance-measuring circuit is called an ohmmeter circuit. There are two types of ohmmeter circuits: the series-ohmmeter circuit and the shunt-ohmmeter circuit. Either one or both may be found in a typical good-quality vom.

The series-ohmmeter circuit includes a source of power (usually a battery), a calibrated meter, a fixed current-limiting resistor, and a variable resistor, as shown in Fig. 2-10A. Of course, there is also a pair of test leads which connect to this resistance-measuring circuit.

For the purpose of discussion here, assume that a 50-microampere meter having a resistance of 5000 ohms is used, that the value of current-limiting resistor R_2 is 22,000 ohms, that R_1 is a 5000-ohm potentiometer adjusted to a value of 3000 ohms, and that the battery is 1.5 volts. With a total of 30,000 ohms, and with 1.5 volts applied to the circuit, if the resistance being measured is zero ohms, the

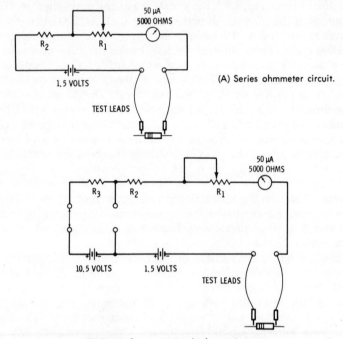

(A) Series ohmmeter circuit.

(B) Circuit for measuring higher resistances.

(C) Shunt ohmmeter for measuring low values of resistances.

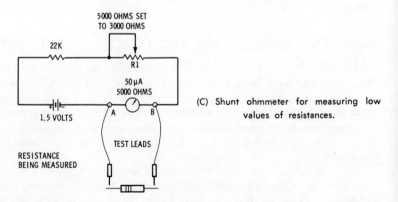

Fig. 2-10. Resistance-measuring circuits.

circuit current should be exactly the 50 microamperes required by the meter to read full scale. In fact, this is essentially how an ohmmeter is calibrated for zero ohms. The tips of the test probes are held together, and the variable resistor is adjusted until the meter reads exactly full scale, corresponding to zero ohms. Now, if the test probes are placed across the terminals of a resistor with a higher value than zero ohms, the deflection of the pointer will not reach the zero-ohms point on the scale. Assume further that the resistance being measured has a value of 30,000 ohms, the same as the measuring circuit. The meter now will be deflected to a ½ full-scale value, since with twice the resistance, the current will be ½ as great. If the probes are placed across the leads of other resistors that have values greater or less than 30,000 ohms, the indication will be either below or above the ½-scale reading of the meter, respectively, by an amount that depends on the difference of the resistance value. Thus, the meter faceplate can be calibrated to read various values of resistance.

Theoretically, this circuit will respond to any value of resistance between 0 and infinity. But, in practice, with the values suggested here, resistance less than 1000 ohms and above 500,000 ohms cannot be measured accurately. This is because these values of about 1000 ohms and 500,000 ohms are very small and very large, respectively, compared to the measuring-circuit total resistance of 30,000 ohms. The 1000-ohm resistor would cause a deflection of about 97% of full scale, and the 500,000-ohm resistor would cause a fairly feeble deflection; neither could be read very accurately.

Circuit for Measuring Higher Resistances

A method of extending the high-end range of an ohmmeter is shown in Fig. 2-10B. A 210,000-ohm resistor (R_3) and a 10.5-volt battery have been switched into the circuit. With the test leads shorted, the current will be:

$$\frac{(10.5 + 1.5)}{(210,000 + 22,000 + 3000 + 5000)} = \frac{12}{240,000} = 50 \text{ microamperes}$$

This is the current required for full-scale deflection. However, at the midrange point on the scale, 240,000 ohms can be read. This is compared to the first example with 30,000 ohms for the 1.5-volt battery and 30,000 ohms of total circuit resistance. Furthermore, assuming accurate reading down to the 10% deflection point of 5 microamperes on the meter, it will be possible to measure resistance up to:

$$\frac{12}{5 \times 10^{-6}} - 240,000 = 2,400,000 - 240,000 = 2,160,000 \text{ ohms}$$

or 2.16 megohms. With the beginning of the 1.5-volt battery and with 30,000 ohms of circuit resistance, the 10% deflection point reading would be:

$$300,000 - 30,000 = 270,000 \text{ ohms}$$

Circuit for Measuring Lower Resistances

It was shown that the circuit in Fig. 2-10A was not satisfactory for measuring low values of resistance. Of course, the circuit in Fig. 2-10B is even less suitable. In commercial ohmmeter circuits, the very low values of resistance are measured by use of the shunt-ohmmeter circuit shown in Fig. 2-10C. The test leads in the measuring circuit now connect across the meter. When the test probes are connected across an unknown value of resistance, the current in the measuring circuit is reduced, part of it going through the meter and part through the unknown resistance. If the resistance is 5000 ohms, the equivalent resistance across points A and B is reduced to 2500 ohms. Although the total current in the circuit increases now to 1.5 divided by 27,000, or about 55 microamperes, the current through the meter itself is half that amount, or about 27.5 microamperes. Thus, measuring a resistance of 5000 ohms would give an indication on the scale at slightly above midrange. Further variations of the shunt-ohmmeter circuit of Fig. 2-10C will permit reading even lower values of resistance. In some vom's, for certain ranges of resistance measurement, a combination of a series circuit and the shunt circuit is employed.

Meter Overload Protection

Most vom's include some means of protecting the meter movement from damage due to overload, or excessive current. In some meters this overload protection is in the form of either a fuse or a circuit breaker which opens the circuit when excessive current flows through the meter. In other instruments the protection is a zener-diode circuit connected across the meter movement. For normal current values, the diode is an open circuit. If the voltage across the meter becomes too high, the zener diode conducts, shunting part of the current around the meter, thus protecting it from damage.

MULTIPLE USE OF SCALES IN A TYPICAL VOM

In a commercial vom the multiplier and shunt resistances for different ranges of voltage, current, and resistance measurement are chosen so that the range-selecting switch can be labeled logically and so that the calibrated scales on the meter faceplate can represent several ranges.

For example, for the vom shown in Fig. 2-11, the calibrated scale second from the top of the faceplate is labeled 0, 50, 100, 150, 200, and 250. This scale is used for the following functions and ranges:

dc volts: 0-2.5; 0-250.
ac volts: 0-250.

The third scale from the top has two separate calibrations; the upper set is labeled 0, 10, 20, 30, 40, and 50; the lower set is labeled 0, 2, 4, 6, 8, and 10. These scales are used for the following functions and ranges:

dc volts: 0-1; 0-10; 0-50; 0-250; 0-500; 0-1000.
ac volts: 0-2.5; 0-10; 0-50; 0-250; 0-500; 0-1000.
dc current: 0-1 mA; 0-10 mA; 0-100 mA; 0-500 mA.

The fourth scale from the top is for use in making low ac voltage measurements, between zero and 2.5 volts; this scale is labeled 0, .5, 1.0, 1.5, 2.0, and 2.5.

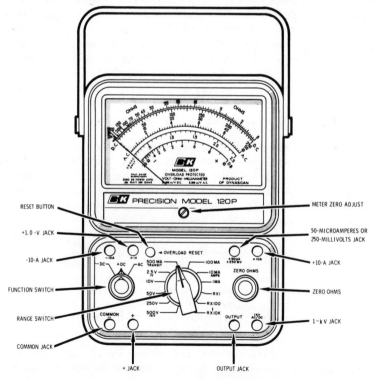

Courtesy B & K Division of Dynascan Corp.

Fig. 2-11. Identification of important features of a typical vom.

The bottom scale is for measuring ac output volts, such as the output signal from an audio amplifier; the value measured is specified in terms of decibels, or dBs. This bottom scale starts from −20 at the left side and progresses toward the right through −10, −4, −2, 0, and then is marked off in positive values of 2, 4, 6, 8, and 10.

The top scale on the meter faceplate of Fig. 2-11 is used for measuring resistance. If the range switch is on the R × 1 position, the resistance values are read directly from the meter. With the range switch in the R × 100 position, the values indicated on the meter are multiplied by 100; and with the range switch in the R × 10K position, the values indicated are multiplied by 10,000. Note that the figure 10 appears at about midrange on the ohms scale. For resistance measurements, if the range switch is set to R × 1, this number 10 indicates 10 ohms; for the R × 100 range, the 10 indicates 10 × 100, or 1000 ohms; and for the R × 10,000 range, the 10 indicates 10 × 10,000, or 100,000.

Note that on this highest resistance range it is possible to read resistances accurately up to the point marked 500, or 500 × 10,000 ohms (5 megohms). Low resistance values can be read on the R × 1 scale to well below 1 ohm.

The round knob at the lower center of the meter is the ZERO OHMS control, equivalent to R_1 in Figs. 2-10A, B, and C. The knob is used for setting the pointer deflection exactly to full scale when the test probes are shorted together.

SENSITIVITY OF A VOM

One of the most important characteristics of a vom is its sensitivity rating. This is specified in terms of *ohms per volt*. The higher the ohms-per-volt rating, the more sensitive the vom. The sensitivity rating is usually shown on the face of the meter or given in the instructional manual. It can also be determined by using one of the following formulas:

$$\text{Sensitivity (ohms per volt)} = \frac{1}{E_{fs}} \times R_m = \frac{1,000,000}{\mu A}$$

where,

E_{fs} is the voltage required for deflecting the meter full scale,

R_m is the resistance of the meter movement,

μA is the current in microamperes required for full-scale deflection of the meter.

The sensitivity may also be determined by dividing the total resistance of the circuit for the range in use by the voltage value of that range. To illustrate, suppose the 0- to 12-volt dc range is in use and that for this position of the range switch there are 240,000 ohms in the measuring circuit. The sensitivity of the measuring circuit is then:

$$\frac{240{,}000}{12} = 20{,}000 \text{ ohms per volt}$$

The same ohms-per-volt rating would apply to all dc ranges of the vom. For the ac ranges, however, the sensitivity is usually rated lower. The lower rating is due to the leakage resistance of the rectifier in the reverse direction and to the fact that the rectified current is not continuous. The ac sensitivity of the vom of Fig. 2-11 is 5000 ohms per volt. Its dc sensitivity is 20,000 ohms per volt. In practice, it is usually important to have a high dc sensitivity rating, the ac sensitivity rating not being quite so important. In many applications, however, a vom having a sensitivity of only 1000 ohms per volt will be satisfactory. A vom of 1000 ohms per volt is much less expensive than one of 20,000 ohms per volt, and the measurements made with the less-expensive instrument will be just as accurate when these voltage measurements are made across resistances or impedances of relatively low value.

Loading Effect of a Meter

The need for using a high-sensitivity vom for voltage measurement in high-impedance circuits becomes apparent if the results are examined when one with low sensitivity is used.

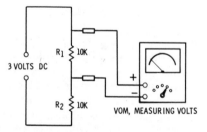

Fig. 2-12. Circuit showing loading effects caused by vom.

For instance, in the circuit of Fig. 2-12, with 3-volts dc applied across two 10,000-ohm resistors (R_1 and R_2) in series, the drop across each resistor is 1.5 volts, and the circuit current is:

$$I = \frac{3}{20{,}000} = 0.00015 \text{ ampere}$$

Now assume the use of the 3-volt range on a 1000-ohms-per-volt meter in order to measure the voltage across R_1. On the 3-volt range, the total internal resistance of the vom is only 3000 ohms.

As soon as the vom test leads are connected across R_1, the 3000-ohm resistance of the meter in parallel with the 10,000 ohms of R_1

has an effective value of approximately 2300 ohms, which changes the total circuit resistance to 10,000 plus 2300, or 12,300 ohms.

The circuit current now becomes:

$$I = \frac{3}{12,300} = 0.000244 \text{ ampere}$$

And the voltage drop across the terminals of R_1 to which the vom leads are connected is:

$$E = 2300 \times 0.000244 = 0.56 \text{ volt}$$

The meter will read this value instead of 1.5 volts which will actually be the voltage across R_1 when the vom is not connected.

The change in circuit conditions caused by connecting a meter to a circuit is called the *loading effect* of the meter. It is apparent that in the example just shown, the loading effect of the 1000-ohms-per-volt meter was considerable. If now a 20,000-ohms-per-volt vom is used to measure the voltage across R_1 in Fig. 2-12, the loading effect is considerably less. On the 3-volt range, a 20,000-ohms-per-volt instrument has an internal resistance of 60,000 ohms. This 60,000 ohms across the 10,000 ohms of R_1 gives a combined resistance of approximately 8,600 ohms. Then, by calculating as before, the voltage across R_1, with the 20,000-ohms-per-volt vom connected, is approximately 1.4 volts. Although this represents an error of about 7%, it is a considerable improvement over the reading of 0.56 volt obtained with the 1000-ohms-per-volt instrument.

Of course, a 1000-ohms-per-volt vom does not have a serious loading effect on every circuit. For instance, the use of the 150-volt range to measure 100 volts across a 1000-ohm resistor would give highly accurate results. The 150-volt range has a resistance of 150,000 ohms. This 150,000 ohms across the 1000 ohms of the circuit under test would have a negligible effect. Of course, a 20,000-ohms-per-volt instrument is not the final answer to measurements in all high-impedance circuits either. If such a vom is used on the 1.5-volt range to measure 1 volt across 100,000 ohms, the loading effect of the 30,000-ohm meter resistance on this range would be quite serious. However, for a high percentage of the measurements in electronics work, the 20,000-ohms-per-volt instrument provides accurate results. For the remaining percentages, either a higher-resistance vom is required or, more often, an electronic meter is employed. The advantage of an electronic meter, such as the vtvm, with respect to circuit loading is considered later in this book.

Higher-Sensitivity VOMs

Volt-ohm-milliammeters having sensitivities of 100,000 ohms per volt, 200,000 ohms per volt, and even higher are also available. In a typical 100,000-ohms-per-volt instrument, the meter movement is rated at 10 microamperes for full-scale deflection. For measuring voltages of 100 volts or more across circuit resistances of 10 megohms or more, a 100,000-ohms-per-volt vom has even less loading effect than the typical electronic meter. In these higher-sensitivity vom's, the meter movement employed is usually the taut-band suspension type.

OUTPUT-MEASUREMENT CIRCUIT

Input and output signals of amplifiers are sometimes measured or specified in watts, sometimes in volts, and sometimes in decibels (dB). Where it is desired to measure in volts, the proper ac range of the vom is used. Assuming that the amplifier is for the audio-frequency range and that the vom has a good response in this range, the reading obtained will indicate the rms, or effective, value of the audio voltage.

If power in watts is the desired measurement, the square of the measured voltage (E) is then divided by the resistance (R) of the input or output circuit. That is:

$$\text{Power in watts} = \frac{E^2}{R}$$

As an example, 90 volts ac is measured across an amplifier output-circuit impedance of 600 ohms. Then the output power may be calculated as:

$$P = \frac{90^2}{600} = \frac{8100}{600} = 13.5 \text{ watts}$$

As in the vom's described previously, most higher-sensitivity vom's have a specially calibrated scale on the meter faceplate. If no dc is present in the ac or audio signal to be measured, one of the ac facilities of the meter may be used to make the measurement in decibels. Ordinarily, however, a vom will have a test-lead jack, such as the second jack from the bottom right in Fig. 2-11, that is labeled OUTPUT. To use this facility, one test lead is plugged into the COMMON jack, and the other is plugged into the OUTPUT jack. Typically, the only difference between the OUTPUT jack and the ac function of the vom is that a capacitor is employed in series with the OUTPUT jack in order to remove any dc from the measurement being made.

In the meter of Fig. 2-11, the value in decibels may be read directly from the dB scale when the range switch is in the 2.5-V position. Higher decibel values may also be read by following directions for the particular instrument being used.

The decibel values are usable directly only for measurements on a 600-ohm circuit. Furthermore, the scale is calibrated on the basis that 0 dB is equal to 1 milliwatt and only, as mentioned earlier, on a 600-ohm circuit. Decibel measurements on circuits other than 600 ohms may be made, but then the readings obtained will only be relative. Charts and graphs are available for converting decibel readings in circuits of various impedances to actual decibel values with respect to 1 milliwatt taken as the 0-dB reference. Often these charts and tables are included in the vom manufacturer's instruction manual.

QUESTIONS

1. What type of meter movement is used in most vom's?
2. What is the purpose of the function switch in a vom?
3. Name the main components of the d'Arsonval meter movement.
4. If a meter movement appears to be defective, and close visual inspection fails to reveal the source of trouble, what should be done to repair the meter movement?
5. What part of the meter movement sometimes is referred to as the armature?
6. If the pointer of a meter movement does not rest at zero when there is no current through the meter, what can be done to "zero" the pointer?
7. Which meter movement would be referred to as being more sensitive, one requiring 50 microamperes for full-scale deflection, or one requiring 100 microamperes?
8. If 40 microamperes causes full-scale deflection of a typical meter movement, how much current will cause half-scale deflection of the pointer?
9. What may happen if considerably more than the rated full-scale current is applied to a meter movement?
10. How can a resistor be used to extend the current-measuring upper limit of a meter?
11. If a meter movement rated at 1 milliampere full scale has a resistance of 500 ohms, determine the value of the shunt required if the movement is to measure 10 milliamperes full scale.
12. In a typical vom, what provision is made for measurement of current on one or more full-scale ranges?
13. What method permits use of a meter movement to measure voltages much higher than the full-scale voltage of the meter?
14. What size of resistor should be used as a multiplier for measuring 1000 volts full scale on a meter movement having 1000-ohms resistance and a full-scale current rating of 1 mA?
15. Since the d'Arsonval meter movement responds only to dc current, how can it be used for measurement of ac voltage and current?

16. Sketch the schematic of a typical meter rectifier circuit.

17. In many vom's, why are the values of multiplier resistors employed in the ac-voltage circuit different from those employed in the dc-voltage circuit?

18. In an ohmmeter, or resistance-measuring circuit, assume that the resistance of the measuring circuit is normally 5000 ohms with the test leads shorted for full-scale indication of the pointer. If the test leads are then connected across a resistor of 5000 ohms, what will be the relative amount of deflection of the pointer?

19. Name the two basic types of ohmmeter circuits.

20. Which of these types is used for measurement of low values of resistance?

21. What is the purpose of the output-measurement facility of a vom?

22. To what does the "sensitivity" rating of a vom refer?

23. What is the sensitivity of a vom that deflects full scale with 50 microamperes through the meter movement?

24. In the usual vom, how does the sensitivity rating for the ac ranges compare with that for the dc ranges?

25. What is meant by "loading effect"?

CHAPTER 3

Inside the VOM

Some basic circuits and features of vom's have been considered so far. Now let's examine the physical makeup of the vom with regard to its components—in other words, types of test leads used, shunt and multiplier resistors, switches and potentiometers, batteries, fuses, and other components. The purpose is to get an idea of what makes up the vom both inside and outside. Many of the components used in vom's will also be found in other instruments to be considered later.

TEST LEADS, PROBES, AND CLIPS

Ordinarily, one pair of test leads is provided by the manufacturer of a vom. Leads about 3 feet long, like those in Fig. 3-1, are common. One has red insulation, and the other has black insulation. Fastened to one end of each test lead is a plug for inserting the lead into the vom jack. Usually the plug on the black test lead is plugged into the jack marked COMMON, NEGATIVE, or MINUS. The plug on the red test lead is inserted into the other jack, which may be labeled OHMS, VOLTS, AMPS, OUTPUT, etc. At the other end of each test lead is usually either a spring-loaded clip (Fig. 3-1A), with jaws for fastening the clip to the circuit or component being tested, or a test probe (Fig. 3-1B) that is held in the hand and touched to the circuit or component. Sometimes the black test lead comes with the clip, and the red lead comes with the test probe. Other special-purpose test leads also are available.

The owner of a vom may want to obtain additional test leads or to replace leads that have become defective. These may be pur-

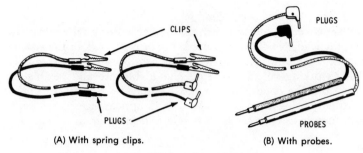

(A) With spring clips.   (B) With probes.

Fig. 3-1. Typical vom test leads.

chased from the vom manufacturer, a local supplier, or a mail-order house. When ordering test leads, the model number of the instrument should be specified since not all test leads are interchangeable, especially with regard to the tips of the plugs inserted in the meter jacks. It is also practical to assemble your own test leads, using the proper wire, plugs, and clips or probes.

Test-Lead Wire

Standard test-lead wire is available from several manufacturers and suppliers. The usual type, as shown in Fig. 3-2A, is about ⅛ inch in diameter. The outer covering is an insulation material; inside this

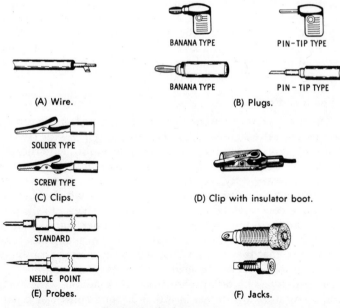

BANANA TYPE   PIN–TIP TYPE

(A) Wire.

BANANA TYPE   PIN – TIP TYPE

(B) Plugs.

SOLDER TYPE

SCREW TYPE

(C) Clips.

(D) Clip with insulator boot.

STANDARD

NEEDLE POINT

(E) Probes.

(F) Jacks.

Fig. 3-2. Test-lead parts.

material are stranded-wire conductors which are spiral-wrapped in a cotton covering for added strength. The wire usually is No. 18, but in some leads No. 22 wire is used. All, or nearly all, of the strands are copper. Standard test-lead wire is generally rated at either 5000 or 10,000 working volts. Test leads should be safe and flexible for easy use, and the conductors should have a low resistance. Therefore, ordinary wire, even if it is similar in appearance, is not satisfactory since it usually tends to tangle and kink, or to be too stiff.

Plugs

Test-lead plugs are made with several types of conductive tips and in several shapes or configurations (Fig. 3-2B). Typically, the plugs are either the pin-tip or the banana-plug type which is larger in diameter and made of springlike sections for added tension and better contact in the vom jack. Most manufacturers of higher-quality instruments provide the banana-plug type.

The body of the plug is generally either a red or a black hard rubber or plastic that covers the tip to protect the user from shock. On preassembled test leads, the two halves of the body may be riveted together; but on plugs purchased for assembly or as replacements, the two halves are usually fastened together with a machine screw and nut that may be removed for connecting the test lead. In other types of plugs, the tip is covered with a barrel-shaped plastic cylinder that may be unscrewed from the tip. On one version of this type of plug, the wire of the test lead must be soldered to the tip terminal; on another, the wire is inserted through the back of the tip, passed through a hole leading out the side of the plug, and fastened by the pressure of a cap nut screwed on the back end of the plug. The most reliable method is to solder the wire to the plug.

Clips and Probes

The probes or clips used at the measuring end of the test leads are also of several different types. The spring types with jaws are generally classified according to shape or assembly as either alligator, crocodile, or meshtooth clips. There are several physical sizes of each of these: heavy-duty, standard, or miniature. Some are provided with protective plastic handles (Fig. 3-2C) or with rubber or plastic protective coverings (Fig. 3-2D) to prevent shocks to the user while he is handling the probes. Both the solder and screw-terminal methods for fastening are available. Also available are clips constructed so that they can be attached to the end of a probe tip, thus making a single set of test leads quite versatile.

Several types of test probes also are available. The types most frequently used are shown in Fig. 3-2E, and consist of a pointed or thin tip fastened to a protective handle of 4 to 6 inches in length, usually colored red or black. The test-lead wire is soldered or fastened by a screw or a pressure nut, depending on the type of probe utilized.

Test leads, probes, plugs, and clips should be inspected regularly for good electrical connection. Look for frayed or loose strands of wire that could cause a shock or a short circuit, and for breaks in the test-lead insulation for the same reason. The proper size test-lead plugs should always be used. Never force plugs that are too large into the jacks (examples shown in Fig. 3-2F) of a vom, or proper contact might be hard to obtain later. Plug tips that are too small should not be used, not only because of the poor contact that might result, but also because the intermittent contact might cause arcing between the tip and the jack. This could result in pitting and perhaps a defective electrical connection when the proper size is later employed.

High-Voltage Test Probe

Many vom's are designed to measure up to 5000 to 6000 volts or more, with the regular test leads provided. No attempt should ever be made to measure voltage any higher than that for which the meter was designed, unless a high-voltage probe is employed (Fig. 3-3). It is better to obtain a high-voltage probe made for your instrument rather than to construct one or adapt one made for another vom.

The voltage-dropping or multiplier resistor contained in the specially designed handle of a high-voltage probe makes it possible to adapt a vom for measuring higher voltages. The resistor value is selected by the manufacturer to match the voltmeter ranges and sensitivity. Instructions accompanying the high-voltage probe list the ranges and multiplying factors for interpreting the readings.

The method of operation of a high-voltage test probe is fairly simple. Suppose a particular meter is rated as 20,000 ohms per volt. Then, on the 3-volt range the vom has a resistance of 60,000 ohms. To measure up to 30,000 volts when the vom is set to the 3-volt range, a total measuring-circuit resistance of $30,000 \times 20,000$, or

Courtesy EICO, Electronic Instrument Co.

Fig. 3-3. High-voltage test probe.

600 megohms, is required. The multiplying resistor in the handle of the probe must then have a value of 600,000,000 − 60,000, or 599,940,000 ohms. When the vom is in use, the common lead would be connected to one side of the high-voltage source, and the tip of the high-voltage probe would be connected to the other side. With the range switch set to 3 volts, the reading is obtained from the 3-volt scale, multiplying the reading obtained by 30,000/3, or 10,000. However, rather than dealing with the cumbersome multiplier 10,000, you could use the 3-volt range setting of the selector switch, read from the 300-volt scale, and multiply the reading by 100. The high-voltage probe can be adapted for use at other high-voltage ranges by employing other settings of the range-selector switch, other scales on the meter, or both.

An important feature of a high-voltage probe is that it is designed to protect the user from shock. The flange located near the upper end of the handle (Fig. 3-3) prevents the user from placing his hand too close to the high-voltage end of the voltage-dropping resistor or too close to the high-voltage circuit being measured. The dropping resistor is made physically long (or consists of several resistors in series) to distribute the voltage gradient over a path as long as possible. This long path helps to prevent arcing in the resistor and reduces the danger of shock to the user.

RESISTORS USED IN VOMS

The typical resistor types used in vom's for multipliers and shunts often have at least 1% accuracy. According to the manufacturing techniques employed, they are known as deposited-carbon, carbon-film, deposited-film, fixed-film, or wire-wound types. Shunt resistors of very low value consist of a piece of copper or iron wire or strap with a resistance that has been very accurately determined.

Of course, should a resistor in a vom become damaged (for example, from excessive current), it should be replaced by one of identical type, preferably obtained from the manufacturer or the manufacturer's distributor.

Practically every vom has at least one potentiometer, that one being used for the ohms zero-adjustment. Potentiometers used in vom's are usually of the carbon type. Some examples of the types of potentiometers found in measuring instruments are shown in Fig. 3-4. The example in Fig. 3-4A has a shaft that is long enough to extend through the instrument panel; the shaft is designed so that a knob may be fastened to the end. The one shown in Fig. 3-4B has a knurled shaft for turning with the fingers; in addition, a slot in the end of the shaft is for screwdriver adjustment. The potentiometer shown in Fig. 3-4C is a printed-circuit type, which mounts to the

(A) Shaft for knob.

(B) Knurled and slotted shaft.

(C) Printed-circuit type.

(D) Wire-wound type.

Fig. 3-4. Potentiometers.

board by means of the tabs on each side and is screwdriver adjustable. Fig. 3-4D shows a wire-wound potentiometer; this type is usually not employed in a vom, especially in the voltage-measuring circuits, because of its inductance which affects the higher-frequency measurements.

SWITCHES

Nearly every vom employs at least one rotary selector switch for changing from one range of voltage, current, or resistance to another. This switch consists of one to four, or more, wafers, with one or two decks of switch contacts per wafer, and two to twelve, or more, contacts per deck. An example of one type of rotary selector switch is shown in Fig. 3-5A.

Sometimes a slide switch, similar to the one in Fig. 3-5B, is employed in a vom. There are several types: single-pole, single-throw (spst); single-pole, double-throw (spdt); double-pole, single-throw (dpst); or double-pole, double-throw (dpdt). Toggle switches (Fig. 3-5C) also are employed with spst, spdt, dpst, or dpdt contacts.

(A) Rotary.

(B) Slide.

(C) Toggle.

Fig. 3-5. Switches used in vom's.

Some vom's have a number of push-button switches for changing functions and ranges.

TYPICAL VOM INTERNAL DETAILS

An internal view of the RCA WV-529A vom in Fig. 3-6 is shown in Fig. 3-7, and the schematic is shown in Fig. 3-8. The appearance and features of some of the components can be identified by comparison of the internal view and the schematic. At this time we will point out a few of the parts shown in the internal view of Fig. 3-7. The dc current shunts are resistors R7, R8, R9, and R10, shown at the lower right. The ac voltage multipliers are R3, R4, R5, and R6 at the bottom center. The dc voltage multipliers at the upper right are R11, R12, R13, and R14. The resistance-measuring circuit includes resistors R17, R18, R19, and R20 at the upper left. Diode CR1 is the rectifier for the ac measuring circuit; diodes CR2 and CR3 provide overload meter protection. The control for ohms adjustment is R24

Courtesy RCA

Fig. 3-6. Example of a vom, RCA Model WV-529A.

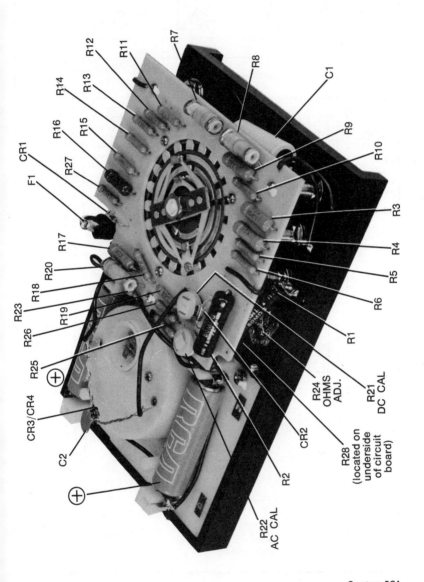

Courtesy RCA

Fig. 3-7. Internal view of the vom shown in Fig. 3-6.

43

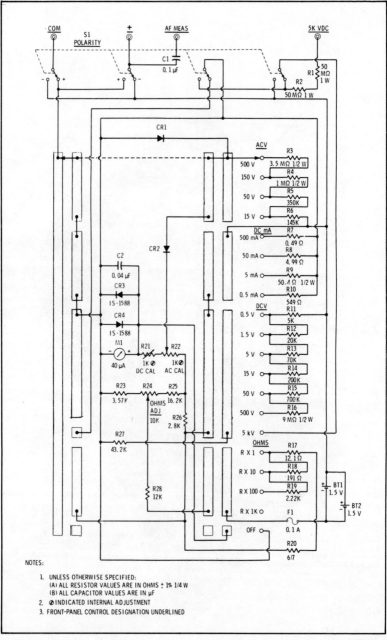

NOTES:

1. UNLESS OTHERWISE SPECIFIED:
   (A) ALL RESISTOR VALUES ARE IN OHMS ± 1% 1/4 W
   (B) ALL CAPACITOR VALUES ARE IN µF
2. ⊘ INDICATED INTERNAL ADJUSTMENT
3. FRONT-PANEL CONTROL DESIGNATION UNDERLINED

Courtesy RCA

Fig. 3-8. WV-529A schematic diagram.

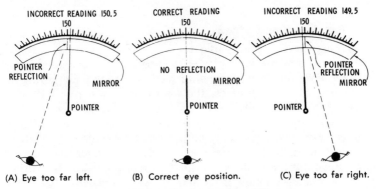

(A) Eye too far left. (B) Correct eye position. (C) Eye too far right.

Fig. 3-9. Use of mirrored scale to avoid parallax error.

at the lower center. The combination range and function switch is the circular device near the right center.

A feature that is included in a relatively high percentage of meters is an antiparallax mirror, shown on the meter face of Fig. 3-6. This mirror makes it possible to read the pointer indication more accurately by eliminating parallax error. This is the error resulting when the eye of the observer is not directly in front of the meter pointer, as shown in Figs. 3-9A and 3-9C. In Fig. 3-9A, with the observer's eye too far to the left, the pointer appears to be at 150.5 volts; in Fig. 3-9C, with the observer's eye too far to the right, the pointer appears to be at 149.5 volts. However, in Fig. 3-9B, the observer knows that his eye is directly in front of the pointer because the pointer and its mirror image coincide and the correct reading, 150 volts in this case, is observed.

VOM ACCESSORIES

Sometimes accessory items may be included with a vom when it is purchased, or the accessories may be obtained separately from the manufacturer. Various types of accessories are available; some make it easier to use the instrument by making measurements more accurately or more conveniently; others extend the range of the instrument. Some examples of these will be considered briefly.

External High-Current Shunts

For most vom's, external shunts are available for extending the range upward in dc current measurement. The shunt usually plugs into the front-panel jacks of the vom and is calibrated for the specific vom to extend a certain range to, say, 30 amperes, 60 amperes, or 120 amperes. External shunts should be obtained from the manufacturer for the exact model of instrument employed.

Carrying Cases and Stands

Carrying cases are available for most vom's; usually these are leather, but some are hard plastic. A carrying case is a good investment for several reasons. Its main function is to provide portability. A carrying case also protects the instrument from physical shock, moisture, dust, dirt, and objects that might strike the vom.

Also available are metal, wire, rubber, or plastic stands that are designed so that the vom may be placed at an angle convenient for use. Most instruments are designed in such a manner that they can either stand upright or lie flat on the bench or other working surface. In the average situation in which measurements are made, it is more convenient for the user if the vom can be tilted backward.

Adaptor Plug-In Units

Some manufacturers make plug-in adaptor units for popular models of vom's. A plug-in adaptor unit, for example, may convert a vom to an audio wattmeter for the installation and service of high-fidelity systems and other audio systems, or for telephones, intercoms, and public-address systems. By means of a switch, the desired load impedance (4, 8, 16, or 600 ohms), may be selected. In the direct position of this switch, normal use of the vom is restored without removal of the adaptor.

Other adaptors can convert a vom to a: transistor tester, temperature tester, ac ammeter, microvolt attenuator, battery tester, milliohmmeter, or extended-range dc ammeter. The main advantage of an adaptor unit (over purchase of a special measuring instrument) is that some savings are obtained since the adaptor does not include a meter movement of its own, instead utilizing the one in the vom.

QUESTIONS

1. What characteristics should the wire used for test leads have?
2. Describe the important features of a high-voltage test probe.
3. If a shunt or multiplier resistor in a vom should become damaged, what should be considered before replacing the resistor?
4. What is the tolerance rating of typical shunts and multipliers in vom's?
5. In Fig. 3-8, which component is the zero-ohms potentiometer?
6. What is parallax error?
7. What device included in some vom's reduces the likelihood that parallax error will occur?
8. What is an external high-current shunt?
9. What should you look for when checking over a pair of test leads?
10. What is the purpose of the flange that is included in the construction of the typical high-voltage test probe?

CHAPTER 4

Putting the VOM
to Work

The specifications for a vom describe the functions and limitations of the vom. Knowing the "specs" for your vom, or one you are about to purchase, is very important in using it effectively.

SPECIFICATIONS AND THEIR MEANING

Some of the specifications and terms important to know include sensitivity, accuracy, and frequency response.

Sensitivity

Sensitivity has previously been discussed to a considerable extent. This specification indicates how many volts, millivolts, milliamperes, or microamperes are required for full-scale deflection of the meter. It has been shown that a meter having a movement rated at 1000 ohms per volt has 1 milliampere flowing through the movement when the pointer is deflected full scale. Also, a 20,000-ohms-per-volt instrument is deflected full scale when 250 millivolts are applied to the movement terminals—at full scale, 50 microamperes are flowing through the movement. The 20,000-ohms-per-volt vom is the most widely used for general electronics servicing work, but other instruments are available with sensitivity ratings of 100,000 ohms per volt, 200,000 ohms per volt, or more.

The sensitivity rating attributed to a vom generally refers to its performance for measuring dc volts. For measuring ac volts the sensitivity generally is lower. For most better-quality vom's the ac sensitivity is 5000 ohms per volt, sometimes 2000 ohms per volt; occasionally, ratings of 1000 or 10,000 ohms per volt (or more) may be encountered, depending on the nature and the design of the rectifier circuits. For the lower ac ranges, sometimes a high ac sensitivity is quite useful, for example, when measuring a low-level input signal to an amplifier stage across a high-value grid resistor.

Insertion Loss

The term "insertion loss" is sometimes applied to a vom and describes its sensitivity or loading effect when it is used for measuring current. For example, when a 20,000-ohms-per-volt vom is used for measuring direct current, the insertion loss is usually specified as 250 millivolts. This 250 millivolts is called a loss because, when the meter is in the circuit, it reduces the voltage applied to the circuit being measured by 250 millivolts when the current deflects the pointer full scale. If the circuit voltage is fairly high compared to 250 millivolts, this loss is negligible. However, if the applied voltage is fairly close in value to 250 millivolts, and if this applied voltage is fixed, the current reading obtained will not be a true indication of the amount of current in the circuit under normal conditions. For example, in Fig. 4-1 the applied voltage is 0.5 volt, or 500 millivolts, and the value of R is 10,000 ohms. By Ohm's law, the current will be $0.5 \div 10,000 = 0.00005$ ampere, or 50 microamperes.

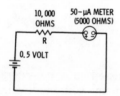

Fig. 4-1. Insertion loss due to presence of current-measuring meter.

For a 20,000-ohms-per-volt meter movement, there will be no shunt when the switch is set for the 50-microampere range, so the resistance contributed to the circuit by the meter will be 250 millivolts/50 microamperes, or 5000 ohms. This 5000 ohms in series with the 10,000 ohms of R brings the circuit resistance to 15,000 ohms. And, under this condition, the current now flowing in the circuit and indicated by the meter will be $0.5 \div 15,000$, or 33 microamperes. The voltage applied to R (the voltage drop across it) will be $0.000033 \times 10,000$, or 0.33 volt. The loss contributed by the meter will be 0.000033×5000, or 0.165 volt, approximately. Thus, in this case, the insertion loss is 165 millivolts. It is only when the circuit conditions are such that the pointer is deflected full scale when the meter is in the circuit that the insertion loss will be the full 250 millivolts specified.

Accuracy of VOMs for Voltage and Current Measurements

The accuracy specified for most vom's is between 2 and 5% for dc voltages, and 2 and 10% for ac voltages. It might be important in some cases to keep in mind exactly how this accuracy factor can affect a reading. Assume that you are using a vom with an accuracy given as ±3% on the dc voltage ranges. This 3% does not mean that any dc reading obtained will be accurate to within 3%. What it does mean is that the reading obtained will be accurate within plus or minus 3% of the maximum value of the range employed.

Suppose, for example, the 3-volt dc range is used. Then, the reading obtained might be higher or lower than its true value by the amount of 3% of 3 volts or $0.03 \times 3 = 0.09$ volt. Thus, if the pointer indicates 2.5 volts, the actual voltage might be either $2.5 - 0.09 = 2.41$ volts, or $2.5 + 0.09 = 2.59$ volts. Similarly, an actual voltage of 2.0 volts might result in a reading any where between 2.09 and 1.91 volts, a possible error of 4.5%. An actual voltage of 0.5 volt might result in a meter reading between 0.59 volt and 0.41 volt, a possible error of 18%.

Thus, on any particular range, the most accurate readings are obtained when the pointer is being deflected as near full scale as possible, and the chance for an erroneous reading increases rapidly for readings of lesser and lesser deflection of the pointer.

A general rule of thumb is that voltage and current readings should be taken in the upper $\frac{2}{3}$ of the scale. Then, when measuring a voltage somewhere in the vicinity of, for example, 0.4 volt, the 0- to 0.5-volt range rather than the 0- to 1-volt or the 0- to 3-volt range should be used.

For the usual good-quality vom, typical accuracies are ±3% for dc ranges, ±5% for ac ranges, and ±3% for current ranges.

Accuracy of VOMs for Resistance Measurement

A vom with an accuracy given as 3% for the dc voltage ranges will have the same basic accuracy for the measurement of resistance. However, the accuracy for resistance measurement must be specified differently; the reason is that at maximum deflection on the resistance scale, the pointer indicates zero ohms, and at minimum deflection the pointer indicates infinity. For this reason the specification usually is given as being within so many degrees of pointer position or within so many percent of the length of the deflection arc.

On many vom's the full arc of deflection between maximum and minimum scale values is 100°. Therefore, on these vom's if the dc voltage accuracy is ±3%, the accuracy on the ohms scale is given as ±3°, or as ±3% of the arc length. Determining from a particular resistance reading what the exact resistance might be (the extreme values it might lie between) is not so simple as determining what the actual voltage might be for a particular reading. However, this is usually a minor matter so long as another general rule of thumb is followed. For resistance measurement, since the resistance scale is crowded toward the high-value end, as in Fig. 4-2, resistance readings should be made in the lower-value half of the resistance scale; the higher (toward zero ohms) on the scale, the greater the accuracy. By staying above the half-way deflection point, the user

Fig. 4-2. Resistance scale of vom showing crowded high-resistance end of scale.

is assured that the readings obtained will be accurate to within 6% for a ±3% scale-length specification. Since it is seldom necessary to measure resistance having a tolerance of less than 5% (most are 10 or 20%), staying within the upper half of the resistance scale will give fairly reliable results.

Interpreting Ohmmeter Scales

As mentioned earlier, the maximum deflection on a vom ohmmeter scale is 0 ohms, and the minimum deflection is equivalent to infinite resistance. The scale of an ohmmeter is labeled progressively from right to left between 0 and some maximum value at which the scale effectively ends. Assume that the ohms scale of a vom is calibrated or labeled for values between 0 ohms and 1000 ohms (1K). In other words, the scale ends at 1K. If the resistance ranges provided were R × 1, R × 100 and R × 100K, the manufacturer of the vom could say that the vom has resistance ranges of 0 to 1000 ohms, 0 to 100,000 ohms, and 0 to 100,000,000 ohms.

When measuring the value of a particular resistor, the proper range to use should be determined by looking at the midscale values for each of the ranges available. This midscale value for each range cannot be determined from the usual ranges specified for a vom. One manufacturer might label his R × 1 range to end at 1K, while another might label his to end at 100K, which would be only a very small distance higher on the scale. The midscale value can be determined, of course, by looking at the ohms scale, but most manufacturers of quality vom's list the midscale values as well as the end-scale values for each range. Therefore, when considering the purchase of a vom, the midscale values given for each range probably will deserve more serious consideration than the end-scale values. A vom that has a midscale figure of 400 ohms is not as useful for measuring low resistance values as one that is 10 ohms at midscale (as is shown in Fig. 4-2).

Frequency Response

Some vom's are designed to be accurate on the ac ranges at 60 hertz (Hz) and, in practice, do not give a reliable indication at frequencies much above or below 60 Hz. Most of the best vom's of the type previously considered are designed for a consistent response throughout the audio range. Manufacturers differ in the way they specify the response of their instruments. The statement "flat from 50 Hz to 50 kHz" is a little indefinite. It might mean flat within ½ dB, within 1 dB, or within 3 dB. However, even if the accuracy were within 3 dB, it could be assumed that this would be a useful instrument for ordinary measurements over the given range.

An example of a method of specifying the response of a vom a little more exactly would be as follows, "±½ dB, 50 Hz to 50 kHz, reference 1000 Hz." This means that the vom is accurate within ±½ dB at any frequency between 50 Hz and 50 kHz as compared to its reading at 1000 Hz. Some manufacturers show frequency response for their instruments in the form of a graph, with separate graphs, if necessary, for each of the ac voltage ranges.

OTHER VOM SPECS

Many other specifications are applied to vom's and other instruments for electrical measurement, especially laboratory instruments. A detailed discussion of all of these would not be in line with the main objective of this book. We will, however, mention briefly three of them: repeatability, tracking, and waveform influence.

Repeatability

Repeatability designates the ability of a vom to repeat readings for successive measurements of the same quantity. Some meters will not give exactly the same reading after the test leads are removed and then reapplied to the same test points. This is due mainly to imbalance or friction in the bearings of the movement.

Tracking

Tracking refers to the ability of a meter to give accurate readings at any point on its scale. For instance, if a voltage is applied so that exactly full-scale deflection results, and then if the voltage is reduced to exactly ½ value, the deflection should be exactly 50% of full scale; similarly, if the voltage is reduced to exactly ¼, the deflection should be exactly ¼, and so on.

Waveform Influence

Nearly all ac vom's are designed for measurement of the rms values of sine waves. However, meters are deflected in proportion to the average value of the sine-wave half-cycles or the average value of whatever waveform is applied. If the waveform is not a sine wave, the value indicated probably will not represent the rms value of the waveform being measured. In circuits where pulses and distorted waveforms are being measured, remember this discrepancy. In cases where it is important to know the exact value and nature of a voltage or current, the oscilloscope should be used.

BEFORE USING THE VOM

Before using a new vom or before using a particular model not used before, the user should examine it closely, and the instruction manual provided for the instrument should be thoroughly studied. Familiarity with the scales, the functions of the switches or controls, and the limitations and advantages of the instrument will be extremely helpful.

ZERO-SETTING THE POINTER

One of the first things to check each time you use a vom is whether or not the pointer is resting exactly on zero. If the pointer does not indicate zero, it can easily be adjusted. First, place the vom in the position (horizontal, vertical, or at an angle) in which you intend to use it. Next, with a thin-blade screwdriver, adjust clockwise or counterclockwise the "zero-set" screw usually located near the center of the vom, just below the faceplate. At the same time you are turning the zero-set screw, gently tap the case of the vom to avoid any slight friction or binding that might prevent the pointer from turning freely. If you do not begin your measurements with the meter pointer exactly at zero, your results will not be accurate.

DC VOLTAGE MEASUREMENTS

In preparing to make a dc voltage measurement, first be sure that the two test leads are in the proper jacks. Usually, the black test lead should be in the COMMON or minus (−) jack; and, in most vom's, the red test lead should be plugged into the plus (+) or

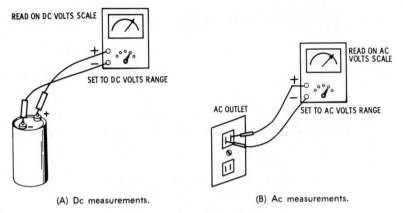

(A) Dc measurements.　　　　(B) Ac measurements.

Fig. 4-3. Dc and ac voltage measurements.

VOLTS jack. If there is a switch marked VOLTS-AMPS-OHMS or AC/DC, be sure that the switch (or switches) is set to VOLTS and DC.

Next, from a schematic or another source, estimate what the voltage to be measured is. Then, with the range switch, select a range that is considerably above this voltage. Turn off the circuit being measured and make sure that no charged capacitors are in the circuit; then connect the test leads across the two points or source of voltage to be measured, as in Fig. 4-3A. Connect the black test lead to the minus side of the circuit and the red test lead to the positive side. Then turn the circuit or equipment on and note the reading. If the pointer appears to be deflecting backwards, either the polarity of the voltage is opposite to what you had assumed, or you have the test leads reversed in the circuit. You must turn off the equipment, reverse the leads, turn it on again, and once more note the reading. If the pointer does not come to rest in the upper ⅔ of the scale, turn the range switch to the next lower voltage position. If the pointer is still below the ⅔ point, move the switch to the next lower setting, and so on. Then, making sure that you are looking at the correct scale for the range you have selected, note the value that the pointer is indicating. For some ranges, the value can be read directly from the scale; for others, it will be necessary to multiply the reading by 10 or 100. For instance, on the 300-volt dc range, if the associated scale is labeled from 0 to 30 and the pointer indicates 25 (Fig. 4-4), you are actually reading 250 volts. Similarly, if the same 0 to 30 scale is used for the 0- to 3-volt range and a reading of 25 is obtained, the voltage is actually 2.5 volts.

In many meters, however, there is a printed scale for each of the ranges provided; thus, multiplying, dividing, or interpreting the readings obtained is seldom necessary.

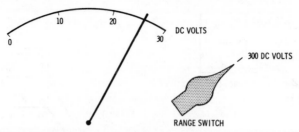

Fig. 4-4. Setting of range switch and position of pointer when measuring 250 volts dc.

Voltage measurements can be made in many cases without turning off the equipment or circuit if the technician has gained sufficient experience to exercise the proper precautions.

AC VOLTAGE MEASUREMENTS

The procedure for making ac voltage measurements is similar to that used for making dc voltage measurements. Begin by making sure the pointer is at zero; plug the black lead into the MINUS jack, the red lead into the PLUS or AC jack; set the AC/DC switch (if there is one) to AC, and set the range switch to an ac range somewhat higher than the rms value of the estimated voltage to be measured. Check to see that the equipment is turned off. Connect the test leads across the points at which the voltage is to be measured, as in Fig. 4-3B; then turn on the equipment and observe the pointer. If there is no deflection or only a little deflection, set the range switch to the next lower range, as required, until the pointer is in the upper ⅔ of the scale.

On many vom's there may be some difference between the ac scales and the dc scales, so make sure that you use the correct scale or scales for ac. It may be that although the positions of the range switch for dc measurements are the same as those for ac measurements, a different set of scales is provided for each on the meter faceplate. On both ranges, the corresponding markings for the high values may substantially coincide; but for the lower values they may not coincide. Also, a completely separate scale may be provided for the 3-volt or other low ac range. After practice it should be easy to select automatically the proper set of scales for ac or dc measurement.

DC CURRENT MEASUREMENT

For measuring dc current with a vom, the circuit in which the current is flowing first must be turned off and then opened, such as at point X in Fig. 4-5A. The range switch of the vom should be set to the current range required. The test leads are then connected in series with the break in the circuit, as shown in Fig. 4-5B, and the equipment turned on. The current is read on the dc voltage scale. If the meter indicates somewhat below ⅔ deflection, the range switch should be set to the next lower current range. If a backward reading is obtained, the test leads should be reversed, or, if a polarity switch is provided, it should be turned to the opposite direction.

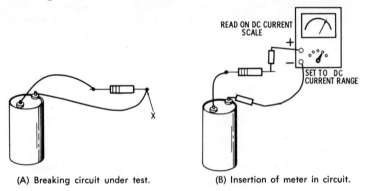

(A) Breaking circuit under test.   (B) Insertion of meter in circuit.

Fig. 4-5. Reading current.

When a vom is set for measuring current, *never* connect the test leads across a live component or source of voltage; this could burn out the meter movement.

As previously mentioned, most vom's do not provide for ac current measurement.

MEASUREMENT OF RESISTANCE

To measure resistance, the range switch is rotated to the correct ohms range, depending on the value of the resistance to be measured. For example, if the ohms midscale value is 5 ohms, and if the estimated value of the resistor being measured is 300 ohms, the range switch should be set to R × 100. If no R × 100 range is provided, the most suitable range is selected to obtain a pointer deflection near midscale.

Before making the resistance measurement, short the tips of the test probes or clips together, and adjust the ohms-zero knob for exactly 0-ohms reading at the extreme right of the scale. Next, connect the test probes across the resistor, as shown in Fig. 4-6. When measuring a resistor in a circuit, at least one end of the resistor should be disconnected from that circuit so that other components in the circuit will not affect the resistance value indicated on the vom. If it is necessary to change the ohms range, the pointer should again be set to zero ohms while touching the probe tips together. This calibrates the range in use and assures greater accuracy.

The same scale is used for all resistance readings, with the scale values multiplied by 1, 10, 100, 1000, 10,000, etc.; these multipliers are determined by the setting of the range switch. On the higher-resistance ranges, touching the ends of the resistor of the test probes with the hands can affect the resistance reading. This is because the body is then connected across the resistor being measured and this parallel resistance lowers the effective value. For high-resistance measurements, touching the resistor or probes should be avoided.

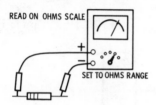

Fig. 4-6. Reading resistance.

One method of preventing this in resistance measurements is to use clips rather than probes to connect to the resistor, thus permitting you to be entirely free of the measurement.

OUTPUT MEASUREMENT

The output-measurement facility of a vom is utilized, for example, in measuring the audio output voltage from an amplifier across a speaker, or in measuring the audio output voltage at the input to an amplifier stage. The output measurement is taken in the same way as an ac measurement, except that, if an ac/dc/output switch is provided on the vom, it should be set to output. This inserts a capacitor in series with one of the test leads, which blocks out any dc present in the circuit that is being measured. The associated ac scale is read for the output value.

Sometimes it is desired to interpret the ac output value obtained in terms of decibels (dB). If the vom includes a dB scale, the value in decibels can be interpreted from that scale. The decibel value depends on the ac range being used. For example, for the vom of Fig. 3-6, to measure decibels, plug the black test lead into the "−" jack; insert the red test-lead plug into the AF jack; set the selector switch to the 15-V range; connect the black test lead to the ground or common of the circuit; and connect the red test probe to the ac voltage test point.

This particular vom has been calibrated so that the value of 0 dB is 1 milliwatt on a 600-ohm line. The decibel values are therefore only relative if the measurement is not on a 600-ohm circuit.

QUESTIONS

1. As applied to a vom, to what does the term "insertion loss" refer?
2. Describe how you would interpret the statement that a particular vom has an insertion loss of 100 millivolts.
3. As a factor that must be considered, is insertion loss of greater significance in low-, medium-, or high-voltage circuits?
4. What is the approximate accuracy of typical vom's?
5. At or near which region of the scale is greatest vom accuracy obtained?
6. If the maximum angle of pointer deflection for a particular vom is 100 degrees, and if the accuracy rating is ±5%, how might the accuracy of the resistance ranges of the vom be specified?
7. If the vom pointer indicates 150 on the resistance scale, and if the range switch is set to R × 100, what is the value of the resistance being measured?
8. How can you easily determine which range to choose when measuring the resistance of a particular resistor?
9. Explain the following example of a specification for a particular vom: ±1 dB, 50 Hz to 100kHz, reference 400 Hz.
10. What does the term "repeatability" mean as a vom specification?
11. What does the term "tracking" mean as a vom specification?
12. What influence does waveform have on the ability of a vom to measure accurately?
13. Explain how to zero-set the pointer before using a vom.
14. Describe how you would set up a vom for measuring dc voltage.
15. Describe how to measure current with the vom.

CHAPTER 5

Use, Repair, and Maintenance

The vom can be used reliably in practically all general testing, troubleshooting, and maintenance situations, except where unusual accuracy is desired or where unusually high-impedance circuits are being tested. The vom, itself, rates fairly high in accuracy, but whenever greater accuracy is required, laboratory or precision instruments must be used. For those circuit measurements where the impedance of the circuit is too high for accurate use of the vom, a vtvm or a solid-state vom is usually employed. However, as mentioned earlier, some vom's are available with input impedances which exceed the input impedance of the typical vtvm for certain voltage ranges. Also, in some applications, digital multimeters offer distinct advantages which will become evident later.

APPLICATIONS IN TESTING AND TROUBLESHOOTING

In the first part of this chapter, general techniques in applying the vom and some of the precautions that should be observed will be considered. The remainder of the chapter will be devoted to methods for the care, repair, and maintenance of the vom.

Measuring Capacitor Leakage Resistance

The vom can be used to advantage in measuring capacitor leakage resistance. For this test, a high-resistance range is employed—for example, the R × 10,000 range. When the ohmmeter leads are applied to the terminals of an uncharged capacitor, the pointer will deflect in the direction of zero resistance, and then either slowly or quickly (depending on the capacitor) the pointer will come to rest at infinity or at a specific amount of resistance. If the capacitor is open, there will be no deflection of the pointer. If the capacitor is shorted, the pointer will indicate zero ohms and remain there. If the leakage resistance is high, the resistance reading will be fairly high compared to the resistance reading obtained from a shorted capacitor.

Generally, the lower the capacitance of the capacitor, the greater the measurable resistance necessary for it to be considered a good capacitor. Mica and paper capacitors of 0.5 μF to 2.0 μF should measure 20 megohms or more. Lower-value capacitors should have an even greater resistance. On the R $\times$ 10,000 range, most capacitors that have a measurement of infinite ohms probably do not have excessive leakage.

Before the leakage resistance of a capacitor is checked, discharge the capacitor by shorting its leads together. A charged, high value capacitor will discharge through the ohmmeter circuit and slam the pointer against the end-stop, damaging the pointer or the movement.

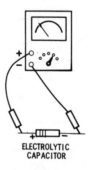

Fig. 5-1. Test-lead polarity for measuring electrolytic-capacitor leakage.

ELECTROLYTIC
CAPACITOR

Electrolytic capacitors will indicate a greater leakage (lower resistance) than paper, mica, or oil-filled capacitors. The ohmmeter test leads must be connected across the capacitor terminals in the proper polarity. The positive ohmmeter lead should be connected to the positive terminal of an electrolytic capacitor, and the negative ohmmeter lead should be connected to the negative terminal, as shown in Fig. 5-1. Connection in the reverse manner usually results in a very low resistance reading.

Because low-value capacitors have a leakage resistance that is not measurable on a vom, the best way to check on them—and on any capacitor, for that matter—is to use a capacitor tester. In determining the leakage of a capacitor, this instrument applies the rated voltage to the capacitor.

Measurement of Capacitance With a VOM

The approximate value of a nonelectrolytic capacitor can be determined by measuring its reactance with the vom and a convenient ac voltage source, such as the power line. The vom, set on the 300-volt ac range, and the capacitor are connected in series and placed across the 110-volt, 60-Hz supply, as shown in Fig. 5-2. The greater the ac voltage reading obtained, the larger the capacitance, assuming that the leakage of the capacitor is low (resistance is high). For a particular vom a table or a graph may be prepared relating the ac voltage reading and capacitance by using as standards several capacitors known to be good. Such a table, provided in the instructional manual for the Triplett 630 vom, is shown in Table 5-1. The ac voltage readings corresponding to various values of capacitance are shown. The same table would also be useful with other similar vom's, provided these instruments had a sensitivity of 5000 ohms per volt on the ac voltage ranges.

It should be mentioned that this method of measuring capacity is only relatively accurate. A better measurement would be obtained by the use of a capacity checker or an LC bridge.

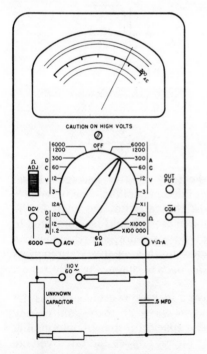

Fig. 5-2. Method of using a vom to measure capacitance.

Courtesy Triplett Electrical Instrument Co.

Forward-Reverse Rectifier Tests

Relative tests on the condition of copper-oxide, selenium, germanium, silicon, or other solid-state rectifiers can be made with the vom. The resistance should measure high in one direction and considerably less in the other direction. With the ohmmeter leads connected across the rectifier terminals so that the lesser amount of current flows (reverse direction), the resistance is approximately 10 times greater than it is with the ohmmeter leads connected across the rectifier terminals in the opposite, or forward direction.

Table 5-1. Relationship of AC Voltage Readings and Capacitance for Measurement Arrangement Shown in Fig. 5-2.

To Measure μF	Set Selector Switch to	Deflection in AC Volts
0.002		0.45
0.004		0.83
0.006	3 volts ac	1.25
0.008		1.65
0.01		2.10
0.02		4.3
0.04	12 volts ac	7.7
0.05		9.7
0.08		14.5
0.1		17.5
0.2	60 volts ac	30.0
0.4		45.0
0.6		57.0
0.8		65.0
1.0		75.0
2.0	300 volts ac	85.0
5.0		95.0
10.0		100.0

In using this method of measurement, care should be taken that low-current signal diodes are not damaged from the normal current of the ohmmeter circuit; with some diodes this method should not be used. Since the forward and reverse resistances of a rectifier depend to some extent on the voltage applied across it, the results obtained should be considered only relative. However, in most instances a defective rectifier shows up as being shorted, open, or having a reverse-to-forward resistance ratio that is too low. Fairly positive proof is possible by measuring a similar rectifier known to be good and then by comparing the readings obtained to those of the suspected rectifier.

Testing Fused Circuits

One of the most useful applications of the vom is in testing fuses and fused circuits. The sources of trouble can be located quickly and easily with the vom. The method of testing a fuse is simple. The power should be turned off, the fuse removed from the circuit, and the test leads of the vom, which is set up as an ohmmeter on the R × 1 range, connected across the fuse. The resistance should be zero, or at most only a fraction of an ohm, for the average fuse in good condition.

Sometimes a fuse opens or develops a high-resistance joint only when the higher current of the circuit in which it is used is flowing through it. In cases when the fuse checks out to be zero ohms by the ohmmeter method but still no power is available to the equipment when it is turned on, a voltage check should be made. This is done, as shown in Fig. 5-3, by setting up the vom for a 120-volt ac reading. One of the test leads is connected to the unfused side of the line, point 1 (Fig. 5-3). The other test lead is then connected to point 2. The vom should read the full line voltage if switch S is on.

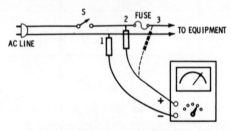

Fig. 5-3. Method for testing fused circuit.

If the full ac reading is not obtained, the trouble is occurring ahead of the fuse, in either the power switch, the ac line cord, the ac plug, or the ac source. Assuming that the proper ac reading is obtained at point 2 (Fig. 5-3), that test lead is then moved to point 3 (Fig. 5-3). If no ac reading is obtained there (or if the ac reading is substantially low), the fuse is probably defective.

Locating Open Filaments

The vom is useful also in locating the open filament in a series-string vacuum-tube circuit. When one tube filament in a series string opens, current is interrupted in all of the tubes in that string. They all go out, and it is impossible to tell by inspection which tube is defective. One method of locating the open tube is to use the ohmmeter section of the vom to measure across each filament; the filament measuring infinite ohms is the open filament.

An other method is to make a voltage check, as shown in Fig. 5-4A. When the vom is set up as an ac voltmeter and is connected across a good filament in a series string that contains an open filament, no voltage reading will be indicated. However, the situation is different if the filament circuit is continuous; in that case, the voltmeter will indicate the rated voltage across the filament. In an open circuit, no current flows through the filament; therefore, there is no voltage drop across it to deflect the meter. However, when the vom is connected across the open filament, the full voltage of the source will be indicated on the meter because the resistance of the vom is much greater than that of the filament. Therefore, the voltage drop across the meter approaches the applied voltage. The connection of the vom across the open filament completes the circuit, permitting a small amount of current to flow and, thus, to deflect the meter. This is not an effective test if two tubes in the string have open filaments. In such a case there will be no voltage reading on any tube in the series string as the vom is moved from tube to tube.

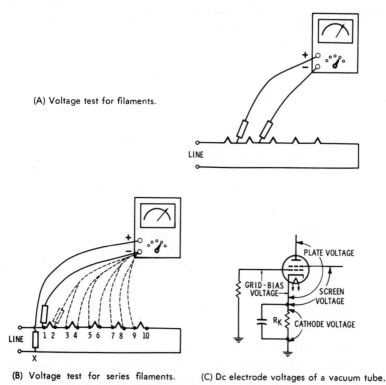

(A) Voltage test for filaments.

LINE

(B) Voltage test for series filaments.

(C) Dc electrode voltages of a vacuum tube.

Fig. 5-4. Measurement of tube-element voltages.

63

A more common and more reliable method is shown in Fig. 5-4B. The meter, set on an ac range that will handle the full line voltage, is connected with one lead to the common side of the line at point X, Fig. 5-4B. The other lead is then moved progressively from point 1 through points 2, 3, 4, 5, 6, etc., getting a slightly lower voltage reading at each tube. If a point is reached where zero voltage is indicated, the tube at that point has an open filament. Replace this tube and proceed to the next filament test points toward the common end of the series string in the same manner. Always work from the beginning of the series toward the end, replacing tubes that have open filaments. The last tube in the string may be checked by the method shown in Fig. 5-4A, since the normal voltage between point 10 and point X is zero.

Testing Electronic Circuits

The dc voltages for the electrodes of a vacuum tube are measured as shown in Fig. 5-4C. Plate voltage is measured between plate and cathode, screen voltage between screen grid and cathode, grid-bias voltage between control grid and cathode, and cathode voltage between cathode and ground (or across the cathode resistor, R_K). Although these are the true dc operating voltages for a vacuum tube, in many cases the voltage lists accompanying the schematic for an electronic device are designated as being measured between the particular tube electrode and the ground or chassis. This reference to ground for each of the voltages is for convenience in measurement; it permits leaving the negative, or common, lead connected to the chassis for each measurement.

Many schematics list the resistance readings between various tube pins and ground. Any serious discrepancy between the resistance listed and the measured resistance is an indication of trouble in one or more of the components common to that circuit.

Quick Check of Transistors

A quick-check method of testing for shorted or open bipolar transistors is shown in Fig. 5-5. This method, which is provided with the RCA WV-529A vom shown earlier, is carried out as follows:

1. Connect the test leads to the transistor elements as shown in the illustration. Set the polarity switch on the meter to POS and measure the base-to-emitter resistance, then the base-to-collector resistance. Both readings should be about the same.
2. Set the polarity switch to NEG and again measure the base-to-emitter and base-to-collector resistances. These two measurements should also be about the same (but different from those obtained in Step 1).

3. The readings obtained should be low (about 500 ohms) for one step and high (about 500K) for the other step, depending upon the polarity of the transistor (npn or pnp). If any of the four individual measurements are radically incorrect, the transistor is bad.

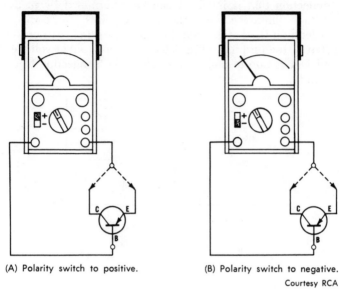

(A) Polarity switch to positive.　　(B) Polarity switch to negative.

Courtesy RCA

Fig. 5-5. Method of quick testing for shorted or open transistors.

Testing Batteries

The vom is useful for testing the condition of batteries. It should always be remembered that it is best to measure the output voltage of a battery when the battery is under load or actually being used in the equipment. As a battery deteriorates or becomes discharged, its internal resistance increases. The load current flowing through this internal resistance reduces the output voltage appearing across the battery terminals. If the load current is not flowing (as when the battery is tested out of the circuit), the voltage is not reduced by the internal resistance and the battery may test good. In many cases, badly deteriorated batteries measure low in voltage even when removed from the circuit. Any battery that measures 75% or less of rated voltage under load is considered weak and should be either charged or replaced, depending on the type of battery.

The vom is very useful for checking the operation of automobile batteries. In many cases, a poor connection due to dirt or corrosion develops between a battery terminal and the clamp or lug fastened to it. When a battery measures full output voltage across its terminals, but poor starting or other electrical troubles are experienced, measure the battery output voltage across the lugs fastened to the battery terminals, and then measure across the terminals themselves. If the voltage measured across both points is not the same, poor electrical contact exists between one terminal and its associated lug. The connection that is involved can be determined by measuring for a voltage difference across the contact points. To do this, place one test lead on the battery terminal and the other lead on the lug connected to the terminal (Fig. 5-6). Any measurable voltage drop is a sufficient reason to remove and clean the connection to correct the difficulty.

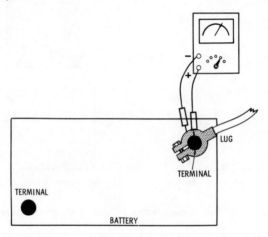

Fig. 5-6. Measuring voltage drop across battery connections.

Measurements in Sensitive Circuits

In some cases, connecting a vom to a sensitive, high-gain or sharply tuned circuit upsets the operation of the circuit being tested so that the reading obtained is not representative of actual operating conditions. Measurement in such circuits is best made with an instrument that has a higher-resistance input. If this equipment is not available, the next best approach is to use a resistor in series with the positive test lead of the vom as shown in Fig. 5-7. The series resistor that is selected should have a value higher than the impedance of the circuit being measured. With the use of the series resistor, the reading will not be as accurate, but often only a relative reading is required.

Where accuracy is important, the reduction due to the series resistor may be calculated. The series resistor and the resistance of the vom on the scale employed may be considered to be a voltage divider. For example, assume that a voltage is to be measured

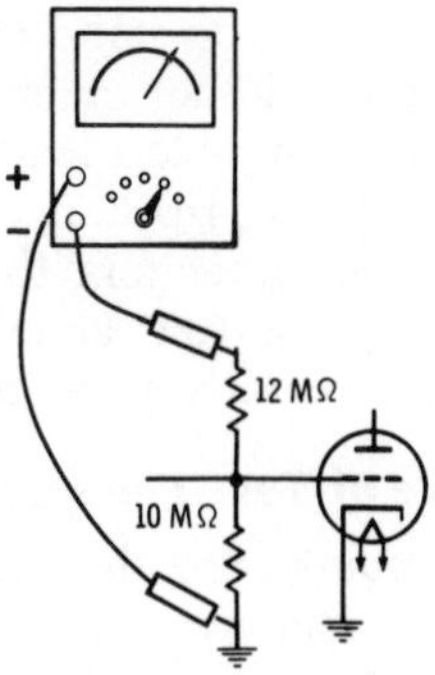

Fig. 5-7. Use of a high-value resistor in series with test lead for sensitive circuits.

across a 10-megohm circuit and that the series resistor selected is 12 megohms. If the 300-volt dc range on a 20,000-ohms-per-volt vom is employed, the meter resistance is then $300 \times 20,000 = 6$ megohms. The voltage divider is then a 12-megohm resistor in series with a 6-megohm resistor, with the measurement occurring across the 6-megohm resistor. With the total resistance being 18 megohms, only $\frac{1}{3}$ of the voltage is read by the meter; therefore, the reading obtained should be multiplied by 3.

The manufacturer of the RCA WV38A vom provides the following table in the instruction manual, listing the multiplier and the reading for specific values of series resistance (Table 5-2).

Table 5-2. Readings Obtained Using Series Resistors to Decrease Loading Effect

Range	Resistor	Reading Multiplier	Scale to Read
2.5 V	50K	2 X	5 V
2.5 V	150K	4 X	10 V
50.0 V	1 MΩ	2 X	100 V
250.0 V	5 MΩ	2 X	500 V

PRECAUTIONS

Caution is essential when working on, or making measurements on, electrical and electronic equipment. You should always be alert to the possibility that the same cause of faulty operation might also cause dangerous high voltage to be present at places least expected.

A good practice is to work with one hand behind you or in your pocket. This gives some protection against contacting points of potential difference. Be sure to avoid standing on conductive, damp, or wet surfaces when making measurements; if possible, stand on

67

a dry board. Try to stand clear of the equipment so that other points on your body do not touch the equipment when you are connecting or disconnecting a test lead.

When making resistance measurements, be sure the power is off and that all capacitors that might hold a charge have been discharged by shorting across their terminals with an insulated test lead or a screwdriver having an insulated handle.

Sequence of Test-Lead Connection

When connecting the vom to a circuit for a voltage measurement, first the common (usually negative) lead should be connected to the chassis of the equipment on which the measurement is being made; then the positive test lead is connected. Fig. 5-8 illustrates what may happen if this practice is ignored. Note that in Fig. 5-8 the positive, or red, test lead is connected to the positive voltage point. If you then happen to hold or touch the tip of the negative test probe and, at the same time, with your other hand touch the chassis (to which the negative lead of the meter is to be connected), practically the full voltage existing across the positive and negative points of the equipment under test will be impressed across your body and the meter. The resistance of the body, normally fairly high, in series with the vom, receives a high percentage of the voltage.

For the reason just mentioned, when test leads are disconnected, the positive, or high-potential, lead should be disconnected first; the negative, or low-potential, lead should be disconnected last.

Determining if a Chassis Is "Hot"

"Hot-chassis" receivers are receivers in which one side of the ac input is connected to the chassis. If the set is connected to the ac

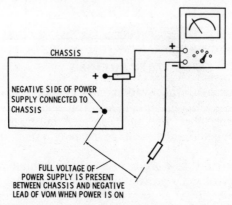

Fig. 5-8. Shock danger of improper sequence of connecting test leads.

output so that the ungrounded side of the ac power line is the one to the receiver chassis, it is possible to receive a dangerous shock by touching the receiver chassis and some ground point at the same time.

The vom can be employed to determine whether or not the receiver chassis is hot relative to ground. Set up the vom to measure 120 volts ac, as shown in Fig. 5-9. Connect one test lead to a ground point, such as a water pipe, radiator, or electric-stove frame, and connect the other test lead to the receiver chassis. If the

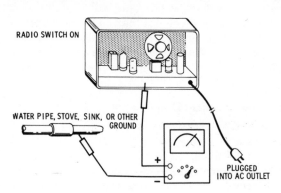

Fig. 5-9. Using a vom to check "hot-chassis" equipment for presence of shock hazard.

vom reads the full line voltage, or even more than a very few volts, the danger of shock exists. To correct this condition, remove the ac power plug from the wall outlet, rotate it 180°, and reinsert it in the wall outlet. Again use the vom to measure between chassis and ground. There should now be no measurable voltage, and the receiver is safer to work on.

Test-Lead Inspection

Test leads should be inspected regularly for broken or frayed leads that could present a shock hazard to the user. Replace or repair such test leads immediately. After a vom test is completed, it should always be disconnected from the circuit immediately; otherwise there is a good possibility you or some other person might unknowingly pick up the vom and test leads and either receive a bad shock or cause a short in the equipment to which the leads are connected.

It is good practice when working on equipment in which more than 40 or 50 volts exist, to have someone nearby in case you do receive a serious shock.

Finally, never make measurements with a vom that has been removed from its case.

CARE AND MAINTENANCE

Manufacturers make vom's as rugged as possible for a delicate instrument, but there is a limit to the abuse an instrument of high sensitivity can withstand.

Care in Selecting Range

It was mentioned earlier, but it is worth repeating, that for current and voltage measurements, always begin with a range higher than the voltage or current expected. This will provide some assurance that the meter will be protected if the voltage or current is excessive. Double-check before connecting the test leads. If the range switch is accidentally left in the OHMS position or on a low range, one or more components of the meter movement (a pretty expensive item) may be destroyed.

Protection From Physical Damage

Always store the vom where it will not fall or accidentally be knocked down; even if the meter cabinet does not break, the pivot of the pointer may be jarred from its bearings. Replacing or repairing any part of the meter movement is usually a job for a specially trained technician. It is usually necessary to return the instrument to the factory for repair and recalibration. Never place the vom on a workbench where power tools are used or where excessive vibration is present. Avoid having the vom where metal chips or metallic dust is present; if these get inside the case of the instrument, a short or other trouble may develop. Do not place the vom where excessively high or low temperatures are likely to occur or where excessive moisture or dampness may cause leakage between components, wires, or switch contacts, or cause deterioration of the batteries.

REMOVAL FROM CASE

On some occasions it is necessary to remove the vom from the case—at least to change the batteries. These get weak after long usage. The first and most obvious sign of aging batteries is that it becomes impossible to bring the pointer to zero in the resistance range, with the test leads shorted together. There is usually some movement of the pointer, even with weak batteries.

For the lower-resistance ranges, usually one or two 1.5-volt dry cells are active in the circuit. For the higher-resistance ranges, such as R × 10,000 or R × 100,000, higher-voltage (but usually not physically larger) batteries of 4.5 volts, 7.5 volts, 33 volts, 67.5 volts, etc., are sometimes utilized. Therefore, it is possible that the vom may zero on one or two of the ranges but not on the others. This evidence will show which battery or batteries to replace.

To remove the vom from its case, it is generally necessary only to remove two to five screws from the back or bottom of the case. These may be either slotted-head screws or Phillips-head screws. In most cases, after the screws are removed, merely lift the case from the instrument. If the back of the case does not come free easily, do not try to shake it free without holding the other part of the case with your other hand.

The batteries usually are held in place by a spring-type holder, and they should not be difficult to remove. All that is usually necessary is to lift them out of the holder and slide in the new battery. Before putting in the new battery or batteries, look at the metal contacts to see if rust or corrosion has started to appear; if so, clean or scrape the contacts. Try to keep the particles from falling into the instrument—if necessary, use a vacuum to remove any that may fall into it.

Batteries should be replaced with similar types. However, for some batteries there are long-life, industrial, or instrument versions for replacement that might give longer satisfactory performance than the type provided with your vom. The leakproof type of battery should be used, which will help prevent the battery chemicals from damaging delicate or precision components. Batteries should be inserted with the proper regard to polarity; the battery holders are almost always marked, one side + and the other −, to correspond to the terminals of the battery.

Fuse Replacement

Should the fuse in the vom blow, it should be replaced only by an identical fuse. If the fuse is one with a conventional element, it should never be replaced by a slow-blow type of fuse; this reduces the margin of protection to the meter movement. A fuse in a typical vom will be a 1-ampere, 250-volt type; fuses of other ratings may also be encountered in some meters.

If the vom fails to respond on all founctions, it is possible that either the fuse is open, one of the test leads is open, or there is a break in the wiring to one of the jacks. The meter movement may be defective if there is no response on any function.

Testing the Meter Movement

If it is suspected that the meter movement is defective, do not attempt to repair it yourself; this almost always leads to added damage to the movement. Instead, follow the manufacturer's directions in the vom instruction manual, and return it to him. Some manufacturers request that you write first, to obtain a "return authorization." Whatever the instructions, follow them closely in order to prevent the meter from becoming lost.

Do not attempt to measure the resistance of a meter movement with another vom, since the batteries in the second vom will probably cause excessive current through the meter.

One way to check the meter movement, if care is used, is as follows: Remove all connections from the terminals on the back of the movement case. Wire a circuit consisting of a 1.5-volt battery and a series resistor for connection to the meter terminals (Fig. 5-10).

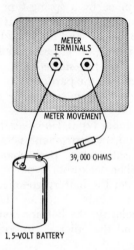

Fig. 5-10. Method for testing meter movement.

Calculate the value necessary for the series resistor to provide about $\frac{2}{3}$ full-scale deflection. For example, if the movement is a 50-micro-ampere, 250-millivolt type, $\frac{2}{3}$ of the full-scale current is about 33 microamperes. The total circuit resistance must be $\dfrac{1.5}{33 \times 10^{-6}}$ = 45,000 ohms (approximately). The meter has a resistance of $\dfrac{0.25}{50 \times 10^{-6}}$ = 5000 ohms. The required value of the series resistor is then $45,000 - 5000 = 40,000$ ohms (39,000 ohms is the nearest standard value). If no deflection is obtained, or if it is substantially different from $\frac{2}{3}$ of full scale, the movement is probably defective.

Rectifier Replacement

Should the meter fail to work properly on the ac ranges, it is likely the rectifier is defective. An exact-replacement rectifier should be used. If two rectifiers are used in the vom, you should replace both, using a properly matched set obtained from the manufacturer or authorized distributor. Substitution of a rectifier that is almost, but not quite, the same will result in inaccuracy on the ac ranges.

Resistor Replacement

If you have the misfortune to try to measure voltage when your vom is set to measure ohms, it is possible that one of the resistors in the ohmmeter circuit will open and will have to be replaced. Be sure to replace the resistor with an identical type. If the identical replacement is not available, and if the vom must be used before one can be obtained from the manufacturer, a resistor of identical characteristics—the same value, tolerance, wattage, and composition —can be used.

Calibration of a VOM

The accuracy of a vom can be checked to a sufficient approximation by measuring a battery known to be good. New dry cells should measure about 1.55 volts per cell. The resistance ranges can be checked by measurement of a precision resistor known to be good. The voms also can be compared with another instrument known to be accurate.

In some vom's little can be done, other than to change parts, if it is discovered that the accuracy is off. Other vom's include calibration adjustments for this purpose. Ordinarily these should not be touched. However, if components are replaced, especially the rectifier, it is sometimes necessary to adjust these calibration controls—closely following the directions of the manufacturer.

Soldering Connections in a VOM

In replacing components or resoldering connections, avoid overheating nearby parts; this may change their value. Use a thin-tipped 25- to 40-watt soldering iron. It is important also to use only rosin-core solder; *never* use acid-core solder.

Many vom's now in use employ printed-circuit boards. Always closely follow the directions of the manufacturer when removing, repairing, or replacing the board and when soldering or unsoldering connections or components.

QUESTIONS

1. How can you use the vom for measurement of capacitor leakage resistance?
2. What is the typical leakage resistance for a paper or a mica capacitor of about 0.5 μF?
3. What is the typical leakage resistance of electrolytic capacitors?
4. How do you measure the forward-reverse resistance of a selenium or a silicon rectifier?
5. How can you use the vom to tell if a fuse is blown?

6. Explain how to use the vom as a voltmeter for locating a tube having an open filament in a series-string circuit.

7. Discuss using the vom for testing battery condition.

8. If connection of a vom upsets the operation of a sensitive, high-gain, sharply tuned circuit, how can this effect be reduced?

9. What are some of the safety precautions that should be followed when measurements are made on electrical and electronic equipment?

10. How can you tell if a chassis is "hot"?

11. How can you check a meter movement to see if it is defective?

12. If a vom works on the dc ranges but not on the ac ranges, what is the likely source of the problem?

13. If the rectifier in a vom is defective, what kind of replacement can be used?

14. Discuss the care that should be used when repairing a vom.

CHAPTER 6

Electronic
Analog Meters

The main difference between the vom and an electronic analog meter, such as the vtvm (vacuum-tube voltmeter), the transistorized voltmeter, or the FET (field-effect transistor) meter, is that the electronic type of vom includes one or more amplifier stages to increase the amplitude of the quantity being measured. There are other differences also; some are advantages, and some are disadvantages. The first type of electronic vom to be considered is the vtvm.

THE VTVM: HOW IT WORKS

An important part of the operating principles of the vtvm and other electronic vom's is the same as that of the standard vom. That is, current flowing through a d'Arsonval meter movement causes deflection of a pointer in proportion to the intensity of the current. Vacuum-tube voltmeters are also much like vom's in appearance. Examples are shown in Figs. 6-1 and 6-2. The unit in Fig. 6-1 is a typical instrument of moderate cost and good performance; the one in Fig. 6-2 is of more elaborate design and is somewhat more costly. Instruments between these two limits of price and performance are representative of meters used widely in measuring, testing, troubleshooting, and experimenting.

In addition to commercially built vtvm's, there are kit types that the purchaser can build from parts provided by the manufacturer. An example is shown in Fig. 6-3. Assembling a kit can save the pur-

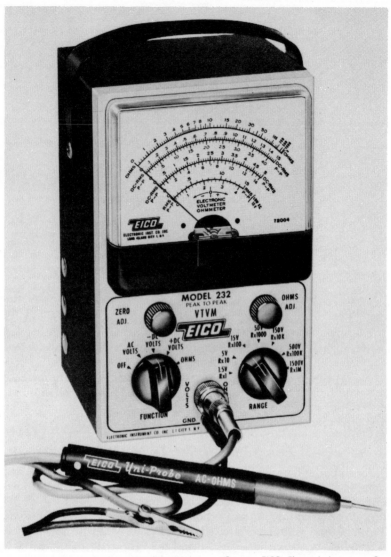

Courtesy EICO, Electronic Instrument Co.

Fig. 6-1. EICO Model 232 vtvm.

chaser a good percentage of the cost of a vtvm if he has the time to
assemble it. Assuming that the directions are followed carefully for
assembly and calibration, and that good soldering practices are employed, a kit vtvm can approach the best commercial models in
performance.

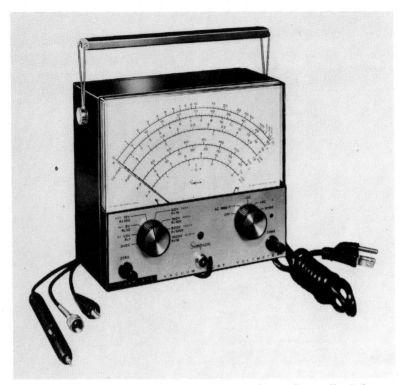

Courtesy Simpson Electric Company

Fig. 6-2. Simpson Model 312 vtvm.

ADVANTAGES AND DISADVANTAGES OF THE VTVM

The major difference between the vom and the vtvm is that in the vtvm one or more vacuum tubes are employed in the circuit. This has the following advantages, as compared with the vom:

1. Higher input resistance.
2. Lower input capacitance.
3. Greater sensitivity.
4. The use of less-sensitive, lower-cost meter movement.

The higher input resistance permits measurement in circuits having high impedance or resistance with less loading effect than with the typical vom. The lower input capacitance of the vtvm makes possible measurement of ac voltage at higher frequencies than are possible with the vom. The greater sensitivity of the vtvm, provided by one or more stages of amplification, makes possible the measurement of lower values of voltage and higher values of resistance. The

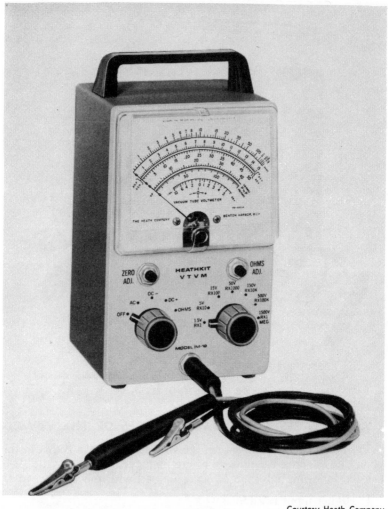

Courtesy Heath Company

Fig. 6-3. Heath Model IM-18 vtvm.

use of the less-sensitive, lower-cost meter movement is made possible by the amplification provided in the vtvm circuit.

These advantages are of sufficient importance, in many cases, to overlook some of the following disadvantages of the vtvm:

1. The vtvm is less stable than the vom; the vtvm requires warmup time for reasonable accuracy.
2. It must be calibrated more frequently.

3. An external source of power is usually required.
4. The more complex circuitry is subject to more-frequent trouble.

The reason for some of these advantages and disadvantages will become apparent later when the basic and typical circuits of vtvm's are discussed.

VTVM PRINCIPLE

Basically, the vtvm consists of an input circuit, an amplifier, and a meter movement, as shown in Fig. 6-4. It is because a vacuum-tube amplifier has a high input resistance that a vtvm causes less loading

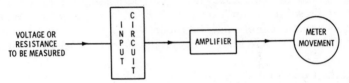

Fig. 6-4. Block diagram of a vtvm circuit.

when it is connected to a circuit for voltage measurement. On most of the voltage ranges, the input resistance for typical vtvm's is 10 or 11 megohms or more.

Simple VTVM

The simplest type of vtvm for measuring dc voltages is shown in Fig. 6-5. The 1-megohm resistor built into the probe is mainly responsible for minimizing the vtvm input capacitance, or capacitive loading effect. It serves to isolate the vtvm circuits from the circuit being

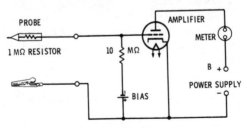

Fig. 6-5. Schematic of a basic vtvm circuit.

measured. The input resistance of this vtvm circuit consists of the 1-megohm probe resistor and the 10-megohm grid resistor, a total of 11 megohms. The battery provides a bias for the triode amplifier tube, keeping it at cutoff until the test leads are placed across a positive or an ac source of voltage.

If the voltage being measured is dc, the positive voltage contacted by the probe lowers the bias on the amplifier grid and causes current to flow through the tube and meter movement in proportion to the amplitude of the positive voltage.

If the voltage being measured is ac, the negative half cycles of the ac voltage have no effect on the amplifier and meter current, since the amplifier is biased at cutoff and the negative ac alternations will increase the bias even further. On positive half cycles, however, amplifier current will flow, the average amount of current causing a proportional deflection of the meter pointer.

It is not practical to use this simple triode circuit in vtvm's, however, mainly because if the voltage to be measured exceeded the bias voltage, the grid would draw current, loading the circuit under test, and resulting in an inaccurate indication on the meter. Another reason is that the probe may be connected only to a positive voltage; this means that there is no provision for measuring negative voltage.

Practical VTVM Circuit

The basic circuit used in many vtvm's is shown in Fig. 6-6. The arrangement in Fig. 6-6A is for measurement of positive voltage. The circuit in Fig. 6-6B (the same as the one in Fig. 6-6A except for the point to which the probe is connected) is for measurement of negative voltage.

The basic vacuum-tube voltmeter circuits of Fig. 6-6 are known as bridge circuits—the meter movement is "bridged" between the plates of two identical vacuum-tube circuits. Suppose no voltage is being measured; the grids of V_1 and V_2 are at the same potential with no grid voltage applied to V_1. Under this condition the currents through the tubes are equal and their plates are at the same potential. With the same potential at each side of the meter, no current flows through the meter, so the pointer indicates zero. If it does not indicate zero, the ZERO ADJUST control is adjusted so that the indication is zero.

VTVM DC VOLTAGE MEASUREMENT

When the test leads in Fig. 6-6A are connected across a source of voltage, with the probe connected to the more positive point, the current through V_1 increases, causing a voltage drop in R_2, and thus decreasing the voltage on the left side of the meter movement. With the right side of the meter now more positive than the left, current flows through the meter, its value being proportional to the voltage applied to the grid of V_1. The current in V_2 does not change, since its grid is grounded. The calibration (CAL) control in series with the

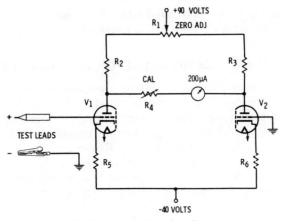

(A) Measuring positive dc voltage.

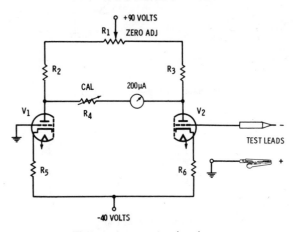

(B) Measuring negative dc voltage.

Fig. 6-6. Practical amplifier circuits for a vtvm.

meter is not an operating control; it is adjusted only at the time of calibration of the meter for exact indication of the pointer.

For measurement of neative voltage, a switching circuit in the vtvm usually transfers the test leads to the opposite triode, V_2, and grounds the grid of triode V_1, as shown in Fig. 6-6B. Now, with a negative voltage on the probe tip, the current in V_2 decreases, the voltage at the right side of the meter increases, and current again flows through the meter in the same direction as that for the circuit in Fig. 6-6A.

As is shown in the schematic, the voltage being measured is applied to the input of each of the vacuum tubes, not to the meter itself. Thus, the meter is isolated from the circuit under test and is relatively safe from damage due to overload.

VTVM MEASUREMENT OF AC VOLTAGE

For the measurement of ac voltage, the same circuit of Fig. 6-6 is used but is preceded by a rectifier circuit (Fig. 6-7A). When ac voltage at the probe swings positive, diode V conducts through resistance R, at the same time charging capacitor C_2 to the peak value of the ac input voltage. Resistor R is of high value, so C_2 does not discharge completely before the next positive half cycle charges it again. The voltage to the grid of the bridge amplifier is approximately equal to the peak value of the ac input voltage.

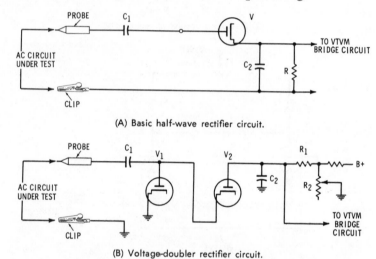

(A) Basic half-wave rectifier circuit.

(B) Voltage-doubler rectifier circuit.

Fig. 6-7. Vtvm rectifier circuits for ac voltage measurement.

Often the rectifier for ac voltage measurement in a vtvm is a twin-diode voltage-doubler rectifier, similar to that shown in Fig. 6-7B. When the ac input voltage goes positive, capacitor C_1 charges through diode V_1 to the peak value of the positive voltage. As the ac voltage swings through zero toward negative, V_1 stops conducting; C_1 remains charged to the peak voltage since it has no discharge path. With the input signal now negative, C_1 discharges through diode V_2 which conducts through C_2. The charging voltage for C_2 is now the sum of the input voltage and that of C_1, or the total of the positive and negative peaks. Thus, the rectifier circuit provides the grid or input of the bridge circuit with a peak-to-peak voltage for the deflection of the meter movement. The scale, however, will be calibrated in terms of rms for a sine-wave voltage and, sometimes, for peak and peak-to-peak values. Potentiometer R_2 permits adjustment for zero deflection of the pointer when a zero-volt input is applied.

VTVM RESISTANCE MEASUREMENT

For measurement of resistance, the input circuit to the vtvm bridge is basically that shown in Fig. 6-8A. When the test leads are shorted together, there will be no deflection of the pointer—a zero-ohms calibration control (not shown here) is adjusted for 0-ohms

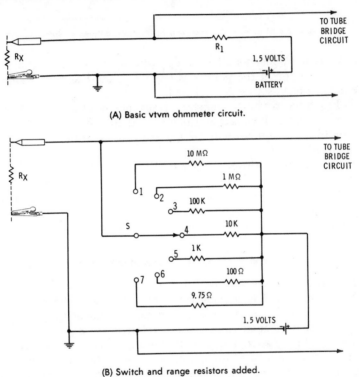

(A) Basic vtvm ohmmeter circuit.

(B) Switch and range resistors added.

Fig. 6-8. Vtvm resistance-measuring circuits.

reading. Then, with the test leads open, the 1.5-volt battery in series with R_1 is across the input circuit, and the meter is deflected full scale (adjusted exactly by means of an OHMS ADJUST control, not shown here). When the test leads are then connected across an unknown resistor, R_x, the deflection of the pointer will be in proportion to the value of R_x. Thus, in the vtvm, as is apparent on the OHMS scale of the vtvm faceplate of Fig. 6-9, the greater the resistance, the

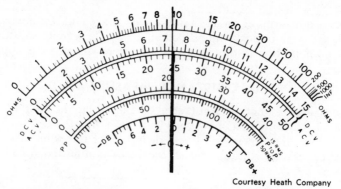

Courtesy Heath Company

Fig. 6-9. Example of vtvm meter faceplate to illustrate differences from the vom.

greater is the deflection. This is opposite to the effect in the vom, where deflection of the pointer is less when the resistance value of the unknown resistance is increased.

In Fig. 6-8A, when the unknown resistance has the same value as R_1, the deflection is midscale, since R_1 and R_x then form a 2:1 voltage divider that applies half the battery voltage to the input circuit of the bridge.

Shown in Fig. 6-8B is the same circuit, but with a switch and additional resistors added for providing seven resistance-measurement ranges. In position 1 of switch S, the midscale reading of the vtvm is 10 megohms; in position 2, the midscale reading is 1 megohm; in position 3 it is 100K, and so on to the lowest range.

VTVM CURRENT MEASUREMENT

The facility to measure current is not usually provided in a vtvm, although some models do include this facility. One of the advantages of the vtvm is that a less-sensitive meter movement may be used; another is that the meter movement is relatively safe from accidental overload. The occasional vtvm that is designed for current measurement cannot usually measure currents as low as the typical vom. The chance for damage to the meter movement increases greatly when the facility for current measurement is added; however, the use of zener diodes and other protective circuits or devices greatly reduces this hazard.

VTVM PROBES

The basic probe for most vtvm's is as previously described. For dc voltage measurement it consists of a housing that contains a 1-megohm resistor in series with the test lead. The 1-megohm series resistor is not used for the measurement of ac voltage or for the measurement of resistance. For these functions, either a different probe is used or a switch built into the probe allows for shorting out the 1-megohm resistor. The circuit of a typical probe is shown in Fig. 6-10A, with the switch in the dc-volts position (1-megohm resistor in the circuit). In the opposite position of the switch, used for ac and ohms, the 1-megohm resistor is shorted out. An assembly view of this

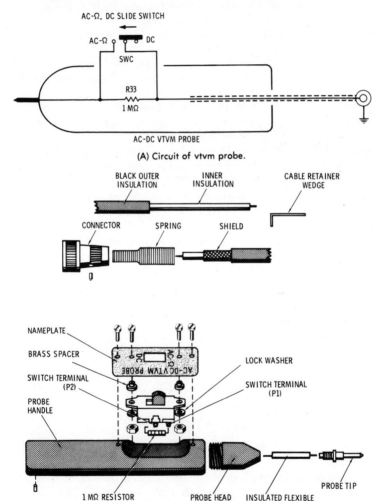

(A) Circuit of vtvm probe.

(B) Assembly view.

Courtesy Precision Division of Dynascan Corp.

Fig. 6-10. Probe for measurement of ac, dc, and ohms.

probe is shown in Fig. 6-10B. A coaxial connector is used for connecting the shielded lead of the probe to the vtvm.

For the measurement of high-frequency ac voltage, an additional probe that is called an rf probe may be utilized with a vtvm. In an rf probe, a diode is built directly into the probe. In this way, capacitive loading from the vtvm on the circuit under test is kept at a minimum; also the vtvm is able to measure a higher frequency range since the output from the probe diode is dc voltage. Therefore, the capacitance of the cable and the vtvm input circuit have no reactive attenuating effect on the signal being measured.

The vtvm, like the vom, can also be used to measure voltages greater than those for which it was basically designed. This is done by means of a high-voltage multiplier probe, the same as the vom high-voltage probe described earlier. Because of the higher average input resistance of the vtvm, a vtvm high-voltage probe has considerably less loading effect on a high-resistance, high-voltage circuit than does the vom high-voltage probe.

For a vtvm having an 11-megohm resistance, the value of the series multiplier resistor, which is in the handle of the high-voltage probe, is 1089 megohms for a 100:1 voltage reduction. The 1089 megohms adds to the 11-megohm input resistance of the vtvm, giving a voltage divider having 1100 megohms of total input resistance. The input to the 11-megohm vtvm measuring circuit is 1/100 of the high voltage being measured. The probe may be used on any of the vtvm voltage ranges where the input resistance is 11 megohms.

RESPONSE OF THE VTVM

The vtvm has a wider frequency response than the vom. A typical vtvm with general-purpose probes provides a flat response within 1 dB or so from 20 or 30 Hz to 3 or 4 MHz or more. With an rf probe, the response can be extended to 250 MHz or more. For some laboratory vtvm's, this response extends to 1000 MHz.

TYPICAL VTVM CIRCUIT

At this point a brief examination of an actual vtvm circuit will help unite the basic concepts previously covered and will give a better overall understanding of the operation of the vtvm. The vtvm we shall consider here is the B & K Precision Model 177 shown in Fig. 6-11; the schematic is shown in Fig. 6-12. This circuit is typical of the popular vtvm's.

The input to the vtvm is provided by means of a common-purpose probe which includes a 1-megohm resistor as previously described.

Courtesy of B & K Precision Division of Dynascan Corp.

Fig. 6-11. B & K Precision Model 177 vtvm.

In the probe circuit (lower-left center), the slide switch is shown in the dc position.

The 100-microampere meter movement, located in the center of the schematic, is connected in the plate circuit of the 12AU7 twin triode in a balanced bridge arrangement. The ZERO ADJ control sets up a balance between the two triodes so that with zero volts applied to the first grid, potentials on each plate are equal. Since there will then be no voltage drop across the meter, the meter will read zero. With a voltage applied to the first grid, the balanced condition is upset, a difference in potentials on the two plates (and across the meter) results, and there will be an indication on the meter. Since there will be a linear relationship between the measured voltage applied to the first grid and the current through the meter, the meter scale is calibrated with linear markings.

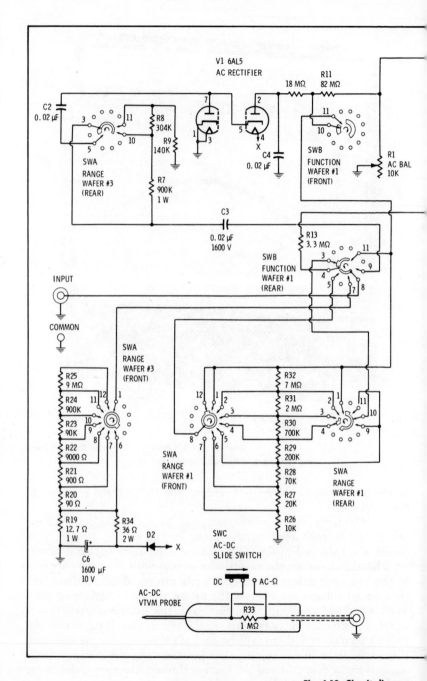

Fig. 6-12. Circuit diagram

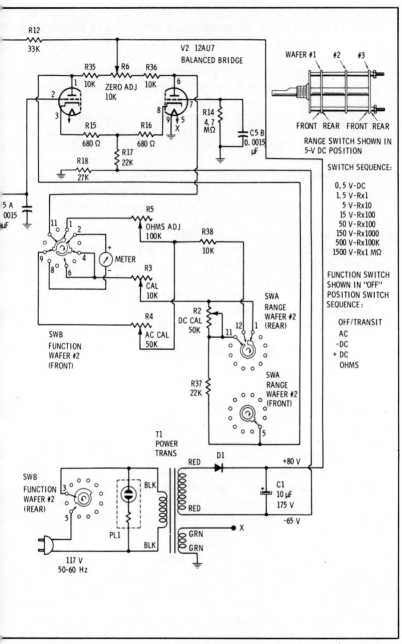

Courtesy of B & K Precision Division of Dynascan Corp.

of B & K Model 117 vtvm.

The maximum voltage ever applied to the 12AU7 is approximately 1.5 volts. The voltage divider at the input, consisting of R21 through R25, has a total resistance of approximately 10 megohms.

For ac measurements, a 6AL5 duodiode is used to rectify the test voltages providing a dc voltage proportional to the applied ac. The dc voltage is then applied through the voltage divider network to the input grid of the 12AU7 balanced bridge circuit, causing the meter to indicate. The 6AL5 is connected as a half-wave doubler which will respond to the peak-to-peak value of applied ac test voltages. The ac voltage scales are calibrated to read both rms and peak-to-peak values. The 0- to 1.5-V and 0- to 5.0-V rms low ac scales have been especially calibrated to improve the accuracy of the meter on these low ranges. Stray pickup reduces the accuracy of any highly sensitive vtvm on the lower ac ranges. In the 0- to 1.5-, 0- to 5, 0- to 50-, and 0- to 150-volt ranges the full ac voltage being measured is applied to the 6AL5 rectifier. A voltage-divider network reduces the voltage on the 0 to 500 and 0 to 1500 voltage ranges to limit the voltage applied to the 6AL5 to a safe level. With proper use of the instrument, the input voltage to the 6AL5 is always 150 volts or less; care should be taken that this value is not exceeded. Developing the habit of starting out with the range switch in the highest position, then working down to the appropriate lower one will result in protection to the instrument. If 400 volts or more is applied to the 6AL5, damage to the tube will probably result.

The ac balance control is used to compensate for the contact potential developed on the 6AL5. A diode tube, with its filament heated, conducts a small amount even though no voltages are applied to plate or cathode. This current flows from cathode to plate of the diode, through the external resistors to ground, and back to the cathode. The voltage developed across the resistors will be negative with respect to ground and is known as *contact potential*. To offset this negative voltage, an equal positive voltage is taken from the power supply and fed into the circuit. The amount of "bucking" voltage is controlled by the ac balance control. This minimizes movement of the pointer when switching from one low ac range to another. The ac calibration control is used to obtain the correct meter deflection for the ac voltage being measured.

For measuring resistances, a 1.5-volt dc supply is connected through a series of multiplier resistances and the external resistance to be measured. This forms a voltage-divider circuit consisting of the 1.5-volt supply in series with one or more multiplier resistors and the resistance under test. The voltage across the unknown resistor is then proportional to its resistance and is applied to the input grid of the 12AU7 balanced bridge circuit which produces an ohmmeter-scale reading proportional to the unknown resistance.

In the OFF position, the meter movement is automatically shorted to prevent damage in transit.

The vtvm is powered by ac voltage for the filaments and by dc voltage derived from the power transformer and the solid-state rectifier, D1. In some older instruments, vacuum-tube rectifiers were employed; most modern instruments now employ either silicon or selenium rectifiers.

QUESTIONS

1. What are the major important differences between the vom and the vtvm?
2. What are some of the disadvantages of the vtvm, compared to the vom?
3. Sketch a schematic of a basic vtvm circuit.
4. Describe the vtvm "bridge" circuit.
5. Of what does a typical vtvm probe consist?
6. What is the purpose of the 1-megohm resistor in the probe of the vtvm?
7. To extend the capability of the vtvm to measure voltages, what type of device is employed?
8. What is the approximate value of the multiplier resistor used in the high-voltage probe for a typical vtvm?
9. By how much does the typical high-voltage probe reduce the voltage applied to the input of the vtvm?
10. How does the rf probe permit measurement of frequencies higher than those that could be measured with an ordinary vtvm probe?

CHAPTER 7

Using and Caring
for the VTVM

In this chapter we will cover some of the practical aspects of using the vtvm, plus some things to keep in mind about care, maintenance, and repair.

The vtvm shown in Fig. 7-1 is similar to the ones we have already discussed except that it is designed to be especially convenient to use on the service bench because it includes a swivel-type bracket, or gimbal, which allows tilting of the vtvm to whatever angle is needed for easy observation. The front-panel controls include a function switch, range switch, zero adjust, and ohms adjust. The probe is a multifunction probe, permitting measurement of ac or dc voltage or resistance by turning the nose of the probe which contains a selector switch.

This vtvm, like most others, operates directly from the ac line. If it is used on a service bench, the technician may wish to plug it in or turn it on when he starts work each day and leave it operating until he finishes. The amount of power consumed is relatively low if left on continuously, and the instrument is always ready for use since recalibration is normally not required.

SAFETY PRECAUTIONS

The following directions for using this vtvm are adapted from the manufacturer's service manual.

CAUTION: It is good practice to observe certain basic rules of operating procedure anytime voltage measurements are to be made. Always handle the test probe by the insulated housing only and do not touch the exposed tip portion.

The metal case of this instrument is connected to the ground of the internal circuit and to the power-line ground through the green line-cord wire. For proper operation, the ground terminal of the instrument should *always* be connected to the ground of the equipment under test. There is always inherent danger in testing electrical

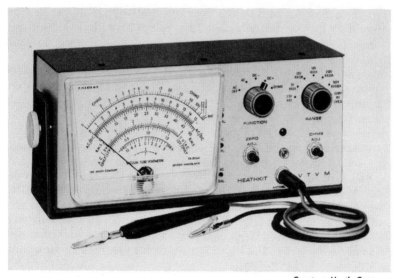

Courtesy Heath Company

Fig. 7-1. Heathkit Model IM-28 "Service Bench" vtvm.

equipment. Therefore, the user should clearly familiarize himself with the equipment under test before working on it, bearing in mind that high voltages may appear at unexpected points in defective equipment.

When measurements are to be made at high-voltage points, it is good practice to remove the operating power before connecting the test leads.

If this is not possible, be particularly careful to avoid accidental contact with nearby objects which could provide a ground return path. When working on high-voltage circuits, play safe. It is a good practice to keep one hand in your pocket to prevent an accidental shock and be sure to stand on a properly insulated floor or floor covering.

COMBINATION PROBE

The probe, which includes the 2-position switch, should be set to AC-OHMS when the FUNCTION switch is on AC or OHMS, and should be set to DC when the FUNCTION switch is on DC+ or DC−. Both the common lead and the probe include a clip so that the user can make measurements without the necessity of having to hold the probe.

UNDERSTANDING THE VTVM SCALE

At the various RANGE switch positions, the voltage markings correspond to the associated full-scale readings. For dc voltage measurements, the corresponding scale is labeled 0, 1, 2, etc., through 15; and 0, 5, 10, 15, etc., through 50. This same scale is used for measuring ac voltages except for the 1.5-volt and the 5-volt ranges. The important features of the vtvm of Fig. 7-1 are shown again in Fig. 7-2 for convenience.

For 1.5 volts dc, read the 15-V scale and move the decimal one place to the left. For example, a reading of 8 would be .8 volt. For 5 volts dc, read the 50-V scale. For example, a reading of 40 would be 4 volts. On the 15-V range, read the 0- to 50-V scale directly. On the 150-V range, read the 0- to 15-V scale and move the decimal one place to the right. For example, a reading of 13 would be 130 volts. On the 500-V range, read the 50-V scale and move the decimal point one place to the right. For example, a reading of 40 would be 400 volts. When using the 1500-V range, use the 15-V scale and move the decimal two places to the right. For example, a reading of 12 would be 1200 volts.

When measuring up to 1.5 volts ac, read the 1.5-V AC ONLY range directly; this scale is lettered in red. On the 5-V range, use the 5-V AC ONLY scale and read the scale directly. This scale is also lettered in red.

Resistance measurements are read on the top scale which is lettered in green. The marking R × 1 indicates that you should read the scale on the Range switch directly. For R × 100, add two zeros to the reading. For R × 10K, add four zeros and on R × 1 MEG add six zeros or read the scale directly in megohms.

The bottom scale, which is marked with zero at the center, is used for making certain tests or measurements where it is convenient to be able to observe the movement of the pointer ither positively or negatively with respect to zero voltage. In using the zero-center feature, the FUNCTION DC− or DC+ and the ZERO ADJ are turned until the pointer is above the "0" mark on the bottom scale, as is shown in Fig. 7-3.

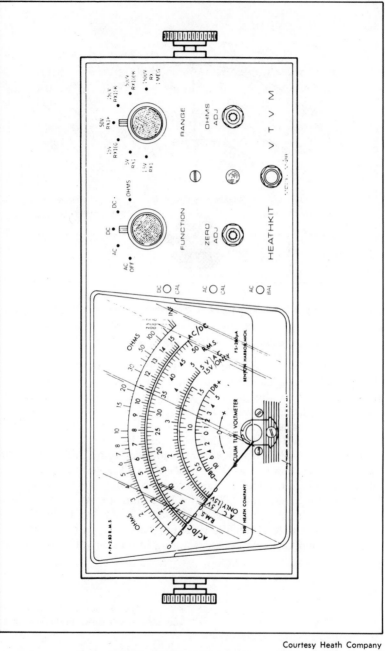

Courtesy Heath Company

Fig. 7-2. Important features on the front panel of the vtvm shown in Fig. 7-1.

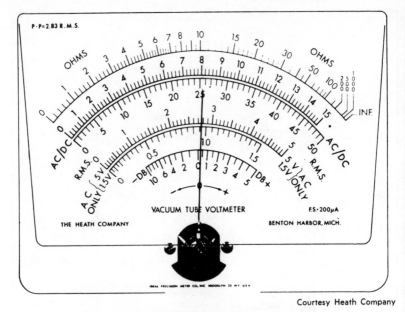

Courtesy Heath Company

Fig. 7-3. Pointer adjusted to zero center on bottom scale.

LOADING EFFECT OF THE VTVM

This vtvm, like most others, has an input impedance of 11 meg-ohms which is considered to be relatively high. The high input impedance is one of the main (if not the major) advantage of the vtvm as compared to the ordinary vom.

Using the manufacturers' examples to illustrate this advantage, assume that a resistance-coupled vacuum-tube type of audio amplifier with a 500K plate load resistor is supplied from a plate source of 100 volts, as is shown in Fig. 7-4. Since the voltage on the plate (between cathode and plate) is shown to be 50 volts, the equivalent resistance of the tube is also 500K. If a 1000-ohms-per-volt vom is used on the 100-volt range to measure this 50 volts on the plate, the resistance of the meter, which is 100K for that range, is placed in parallel with the 500K of the tube, resulting in an equivalent resistance

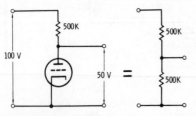

Fig. 7-4. Typical vacuum-tube circuit and equivalent circuit shown to the right of it.

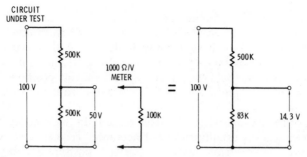

Fig. 7-5. Equivalent circuit of Fig. 7-4 before a 100 Ω/V vom is connected to it (left) and the voltage drop after the vom is connected (right).

of 83K. The voltage on the plate of the tube will then drop to 14.3 volts, as is shown in Fig. 7-5. This large error is caused by the shunt resistance, or "loading effect," of the meter. The fact that this low-resistance meter is being used to measure voltage in a high-impedance circuit changes the conditions in the circuit so that the indication on the meter does not accurately represent the circuit voltage when the meter is not connected.

Fig. 7-6 shows the same circuit but shows the 11-megohm resistance of the vtvm as being connected across the 500K tube resistance. The 11-megohm resistance of the vtvm in parallel with the 500K of the tube has an equivalent resistance of 480K which is relatively close to the original 500K, so the vtvm has relatively little "loading effect" on the circuit. The voltage on the plate of the tube will be reduced by only 1 volt, to 49 volts, as shown.

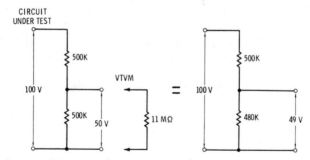

Fig. 7-6. Same conditions as in Fig. 7-5, except using a vtvm to measure the voltage.

MEASUREMENTS

Measuring DC Voltages

To measure positive dc voltages, connect the common or black test lead to the cold, or common, side of the voltage. In equipment that is operated from the ac line through a power transformer, common is usually the chassis. Set the FUNCTION switch to DC+.

Set the RANGE switch sufficiently high so that the voltage to be measured does not exceed the range setting. If the voltage is unknown, set the RANGE switch first to the 1500-volt position.

Touch the test probe (DC position) to the voltage point. If the meter does not read in the upper ⅔ of the meter scale, reduce the setting of the RANGE switch. A meter reading in the upper portion of the meter scale is usually the most accurate. To measure −dc voltages, turn the FUNCTION switch to the DC− position and repeat the previous steps.

Power-Line Measurements

The manufacturer's directions for measuring ac voltages of the vtvm shown in Fig. 7-1 are adapted for the purposes of this discussion as follows:

WARNING: When your power-line outlet is the 3-wire, polarized type, *do not* use the common (negative) lead of this vtvm to measure power-line voltages. To do so may short circuit the power line through the common lead, the chassis, and the green line-cord wire.

1. Set the FUNCTION switch to AC, the range switch to 150 v, and the meter probe to AC.
2. Move the meter common lead out of the way, as it will not be used.
3. Touch the meter probe to one side of the power line. If there is no indication on the meter, you have selected the common side of the ac line; touch the probe to the other side of the line.
4. To obtain contact to a wall outlet, insert a screwdriver blade into one of the outlet openings and touch the probe to the exposed part of the screwdriver blade. Try both outlet openings. *Be careful.*
5. If you have occasion to measure a 240-volt outlet, such as for an electric range or a dryer, you will get voltage readings with the probe at two of the three openings. Add these readings together to get the actual value of the voltage present.

Measurement of Other AC Voltages

To measure other ac voltages with the vtvm, connect the common (black) lead to the common, or "cold," side of the voltage to be measured. Set the FUNCTION switch to AC and set the RANGE switch to a range greater than the voltage to be measured, if known. If unknown, set it to 1500 V. With the test probe in the AC position, touch the point in the circuit at which the voltage is to be measured. If the meter moves less than $\frac{1}{3}$ of full scale, switch to the next lower range. The maximum ac voltage that can be safely measured with your vtvm is 1500 volts, and this limit must not be exceeded. The meter scale of the vtvm is calibrated in rms.

The ac voltage readings are obtained by rectifying the ac voltage and applying the resulting dc voltage to the vtvm circuitry. The rectifier circuit is a half-wave doubler, and the dc output is proportional to the peak-to-peak value of the applied ac.

For sine-wave voltages (Fig. 7-7A), the rms value is 0.35 times the peak-to-peak value. For complex waveforms (Fig. 7-7B) this ratio does not necessarily hold true and may vary from practically zero for thin spikes to 0.5 for square waves (Fig. 7-7C). For sine-wave voltages over 5 volts, the rms value is read on the same scale as a dc voltage. When you are using the 1.5- and 5-volt ranges, the 1.5- and 5-volt ac scales should be read.

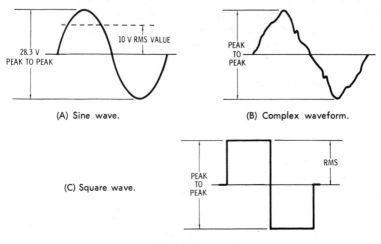

(A) Sine wave.

(B) Complex waveform.

(C) Square wave.

Fig. 7-7. Ac waveforms.

When connecting the vtvm to the circuit under test, the vtvm input resistance R and input capacitance C are effectively placed in parallel with the voltage source. This may change the actual voltage to be measured through loading.

At low frequencies, such as the power-line frequencies of 50 or 60 Hz, the effects of capacitance loading may usually be disregarded; thus, the loading by the vtvm may be considered the same as connecting a 1-megohm resistor across the voltage source. At higher frequencies, the capacitive reactance decreases. At 10 kHz, for example, it is approximately 170K. Such a value may seriously affect the voltage at the point of measurement.

The loading effect of both input capacitance and resistance depends on the source impedance. In low-impedance circuits, such as 50 to 600 Ω, no noticeable error is introduced in the voltage reading through circuit loading. Then the specified frequency response of the vtvm becomes the limiting factor.

As a general rule, it should be kept in mind that frequency response and loading may affect the accuracy of the voltage reading obtained. Consider the resistive loading of 1 megohm regardless of frequency, and the capacitive loading effect at the frequency involved. The actual capacitance of the instrument and leads may also affect the tuning of low-capacitance resonant circuits.

Knowledge of the values in the circuit under test and the values of the input R and C of the vtvm will permit valid readings to be obtained for a wide range of impedances within the full frequency response of the instrument.

Since the vtvm is a sensitive electronic ac voltmeter and since the human body picks up ac when near ac wires, the meter will indicate this pickup. Never touch the probe when the vtvm is set on the lower ac ranges. Zero should be set with the probe shorted to the common clip.

Resistance Measurements

To measure resistance with the vtvm, connect the common (black) lead to one side of the resistor or circuit to be measured. Set the FUNCTION switch to OHMS and set the RANGE switch to such a range that the reading will fall as near midscale as possible. Set the OHMS ADJ control so that the meter indicates exactly full scale (infinity on ohms scale) with the test lead (AC position) not connected to a resistor or circuit. Then touch the test probe to the other side of the resistor or circuit to be measured. Read the resistance on the OHMS scale and multiply by the proper factor as shown on the RANGE switch settings.

Although a battery is used to measure resistance, the indication is obtained through the electronic meter circuit; therefore, the vtvm must be connected to the ac power line and turned on. Establish the habit of *never* leaving the instrument set in the OHMS position because this could greatly shorten the life of the ohmmeter battery, particularly if the test leads are accidentally shorted together when lying on the service bench.

Using the VTVM Decibel Scale

Different reference points for 0 dB have been adopted for various purposes. For audio work, the reference for 0 dB is usually standardized as being 1 milliwatt in a 600-ohm load. The voltage across the load can be calculated to be 0.774 volt.

On the 0- to 15-volt scale of the vtvm being discussed, 0 dB will be 7.74 volts. When an ac range lower or higher than the 0- to 15-V range is used, 10 dB should be subtracted from or added to the value indicated for each change in range position. For example, the indication of 0 dB when the RANGE switch is set to 50 V means the level is actually +10 dB; or an indication of 0 dB when the range switch is set to the 5-V range means the level is −10 dB. When the reference level of 1 milliwatt (mW) in 600 ohms is the standard for 0 dB, this level is usually referred to as 0 dBm. The graph shown in Fig. 7-8 makes it convenient to convert ac voltage readings across 600 ohms to decibel readings, assuming that the 0-dB reference is 1 mW.

As an example of how to use this graph, if you measure 10 ac rms volts across a resistance or line of 600 ohms, the equivalent value will be 22 dBm; similarly, 2 ac rms volts across 600 ohms will be equal to 8 dBm, etc.

For circuits other than 600 ohms, Table 7-1 may be employed to determine the correction factor. For instance, if we measure 20 ac rms volts, the chart shows this to be equal to about 28 dBm for a 600-ohm circuit; but if the measurement is made across a 150-ohm circuit, the table indicates that we must add 6 dB—the actual value then will be 28 + 6, or 34 dBm.

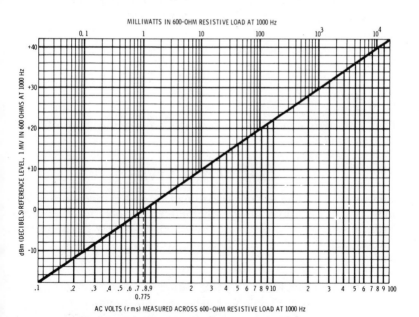

Fig. 7-8. Graph relating dBm to ac rms voltages.

For circuits other than 600 ohms, and not shown in the table, the following formula may be used. In the formula, R represents the resistance of the circuit in which the measurement is made. If R is higher than 600 ohms, the correction factor is negative. For instance, suppose we measure 9 ac rms volts across 6000 ohms. From the

Table 7-1. Correction Factor, $10 \log \dfrac{600}{R}$

Resistive Load at 1000 Hz	dB*
600	0
500	+0.8
300	+3.0
250	+3.8
150	+6.0
50	+10.8
15	+16.0
8	+18.8
3.2	+22.7

* dB is the increment to be added algebraically to the dBm value read from Fig. 7-8.

graph of Fig. 7-8, it is noted that 9 ac rms volts in a 600-ohm circuit is 21 dBm. The correction factor which must be added (subtracted, in this case, since 6000 is higher than 600) is:

$$\text{Correction factor} = 10 \log \frac{600}{6000}$$
$$= 10 \log 0.1 = 10 \times (-)$$
$$= -10 \text{ dB}$$

Thus, the dBm equivalent of 9 ac rms volts across 6000 ohms is 21 − 10, or 11 dBm.

CARE OF THE VTVM

In general, the suggestions regarding care and maintenance of the vom or nonelectronic analog multimeters, which were covered earlier in this book, also apply to the vtvm. To review briefly for application to the vtvm, the following precautions should be observed:

1. The vtvm should be properly calibrated before using it.
2. Batteries, when weak, or leaking, should be replaced.
3. Precautions regarding safety should be followed—connect the common lead first, be careful of hot-chassis equipment.

4. Test leads should be kept in repair or not used if defective.
5. A safe storage location should be provided; locations near machinery, dust, dirt, excessive temperatures and humidity should be avoided.
6. Only replacement fuses and overload diodes of the same rating should be used, and replacement parts of the same characteristics and rating should be employed.
7. Repair of the meter movement should not be attempted.

In the next paragraphs, additional items will be considered with regard to the use and care of the vtvm.

CAUSES OF FAILURE OR INTERMITTENT OPERATION

Some causes of troubles in vtvm's are due to negligent or accidental misuse of the instrument. When you first notice something wrong with the vtvm, think back to the last time you used it for that particular function or range. You may possibly remember what you might have done wrong and will then be able to make repairs more quickly.

For example, perhaps you had the function switch set for ohms when you were attempting to measure voltage. Then, noting that there was no expected voltage reading, you double-checked the switch settings, noticed that the function switch was not set to the voltage positions as required, corrected that situation, and then proceeded with the voltage measurement, not realizing that you had burned out a resistor in the ohmmeter circuit. In an accident of this sort, usually it is one of the low-value, ohms-multiplier resistors that is damaged. Typically these have values of 9.45 or 95 ohms, although the values may vary in different instruments.

Many troubles are also due to failure of a component, tube, cable, switch, control, or the wiring. If trouble occurs and you do not know where to start, the best approach is first to determine just which functions of the vtvm do not respond properly. Usually you can isolate the trouble to one or two major circuits. Begin by inspecting those parts and comparing the voltages, resistances, etc., with the normal operating values provided in the manufacturer's instruction manual for the particular instrument being used.

Tube Failure

Occasionally a tube will fail in a vtvm. The result will be noticeable in the operation. The vtvm may fail to respond at all, may not calibrate properly, may respond only on the ac voltage readings, may operate intermittently, or may operate properly only after an unusually long warmup period.

If you are unable to balance the vtvm, or if the balance is unstable, the cause may be the double triode (12AU7, ECC82, etc.) or the twin diode (6AL5). If there is no reading on any of the ac scales but dc readings are normal, the usual cause is a defective 6AL5. When tubes are replaced, the vtvm should be recalibrated as described in the instrument manual. Before final calibration, the tubes should be aged in the circuit, since their characteristics may change during the first few hours of operation. A tube can be pre-aged if you are anticipating a failure. Preaging can be done by applying 100 to 125 volts dc on the plate, connecting the grids and cathode together and to the −dc source, connecting the filament to its rated voltage, and operating the tube in this manner for 40 to 50 hours. Then place the tube in the vtvm, and recalibrate according to the manufacturer's instructions.

Inaccurate Resistance Readings

If the resistance readings are inaccurate, it may be because the 1.5-volt battery is weak. A way to test the battery will be described later. If the battery does not prove to be weak, one or more resistors in the multiplier circuit may be open or off value—one symptom of a defective resistor is that resistance values measured on the lower ohms ranges will creep in value—the reading will vary as you watch the pointer.

Other Causes of Failure of the VTVM

Other causes of failure of the vtvm include a blown fuse, a faulty on-off switch, an open line cord, a break in the wiring, a poor shield connection on the probe cable, a low dc power-supply voltage, poor socket contacts, a damaged probe resistor, a defective control, or wiring shorted to the metal case.

Battery Test

To ensure accuracy of resistance measurements, the battery should be occasionally tested as follows:

1. Turn the FUNCTION selector to OHMS. Set the RANGE control to $R \times 1$ position.
2. Rotate the OHMS ADJUST control for full-scale deflection of the pointer. Short the probe to the ground clip for about ten seconds.
3. Open the circuit and observe the indication. Any appreciable deviation from full-scale deflection indicates weak cells that should be replaced.

The reason the pointer may fail to deflect full scale is that with the prolonged short circuiting of the probe tip to the ground clip, a fairly high current demand is placed on the battery. This lowers the voltage of a weak battery so that it does not permit immediate full-scale deflection of the pointer when the short is removed.

SPECIAL APPLICATIONS

On many vtvm's, the highest calibrated number on the ohms scale is 1K. If you are reading ohms and have the range switch on the highest setting (1 megohm), the highest readable resistance is 1000 megohms, which is high enough for most purposes. However, on some rare occasions or for special purposes, it may be necessary to measure a higher resistance. There is a way to do it using an external battery and series resistor, as shown in Fig. 7-9.

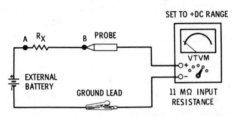

Fig. 7-9. Circuit showing the use of a vtvm to measure resistance beyond 1000 megohms.

The method described by the manufacturer of the RCA WV-98C, which could apply directly or with slight modification to any comparable vtvm, is as follows. The battery symbol may represent an external voltage source of from 20 to 500 dc volts, whatever is required to make a deflection of the pointer. Connect the circuit as shown and then:

1. Set the function selector to +DC volts and measure the voltage at point B (Fig. 7-9).
2. Measure the voltage at point A (Fig. 7-9).
3. Compute the unknown resistance, R_x, from the following formula:

$$R_x \text{ (megohms)} = \frac{11 \text{ (volts at A)} - \text{(volts at B)}}{\text{(volts at B)}}$$

Example: The value of an unknown resistance is to be determined with the circuit of Fig. 7-9. An external voltage of 500 volts is applied. The WV-98C measures 2.5 volts at point B and 500 volts at point A. Then,

$$R_x = \frac{11 \text{ (500} - 2.5)}{2.5} = 2200 \text{ megohms (approximately)}.$$

Peak-to-Peak Voltage Measurement

The vtvm can be used to measure peak-to-peak voltages by reading directly from the peak-to-peak scales of the vtvm. In most cases, even for complex waves and other nonsinusoidal waves, the readings that are obtained will be accurate. In some cases, where the voltage wave being measured consists of pulses of very short duration, or pulses between which there is a long interval, the peak-to-peak voltage reading obtained will be lower than the actual value which would be more accurately indicated when using an oscilloscope.

High-Frequency Measurement

The vtvm's previously discussed have responses up to 3 to 4 MHz and down to 30 to 40 Hz. For most vtvm's this response is accurate when the measurement is being made across a specific value of resistance such as 100 ohms, 600 ohms, 1000 ohms, etc. For measurement across a circuit of some other resistance, the response may differ.

For example, a certain vtvm that is flat to 4 MHz when the measurement is made across 100 ohms may be flat only to 500 kHz when the measurement is made across 1000 ohms. This possible deviation in response for circuits of different resistance should be remembered when making frequency-response checks on audio- and video-amplifier circuits.

For making measurements of voltages having frequencies above the specified response of the vtvm, the crystal-diode probe (available as an accessory for most vtvm's) should be used. This will extend the response of the vtvm to 250 MHz or more, depending on the vtvm and its associated probe. A crystal-diode probe sometimes is an additional cable and probe that must be used instead of the ac/ohms/dc probe or must be inserted in a different jack. In other cases, the crystal-diode probe is simply placed over the standard probe and used directly.

ERRATIC VTVM READINGS DUE TO STATIC CHARGE

Those vtvm's (and vom's) having plastic covers on the meter face may accumulate a charge of electricity when the cover is polished or cleaned. This may cause the pointer to deflect erratically whether the instrument is on or off. The static charges may easily be removed by using one of the commercially available antistatic solutions or a solution of any good liquid detergent and water. Dip a clean, soft cloth in the solution and wipe the surface of the meter cover. The cover need not be removed for this operation.

QUESTIONS

1. How do some vtvm's provide for low-voltage ac ranges?
2. What type of rectifier is commonly used in the power supplies of vtvm's?
3. What is the purpose of the TRANSIT position found on the function switches of some vtvm's?
4. Why does the manufacturer of a vtvm place the calibration controls inside the case?
5. How is the pointer of the meter in a vtvm set to zero?
6. How long should a vtvm be allowed to warm up before use?
7. What precaution should be taken before the ohms function of a vtvm is calibrated?
8. What precaution should be taken when a tube in a vtvm is replaced?
9. What is the purpose of a zero-center scale on a vtvm?
10. What reference is used when decibel measurements are made with a vtvm?
11. Discuss generally what sort of care the vtvm requires.
12. What are some of the causes of failure of vtvm's?
13. What should be done after a tube in a vtvm is replaced?
14. What might be causing the trouble if resistance readings on the vtvm creep in value as you watch the pointer?
15. How can the vtvm be used for measurement of resistances considerably higher than the highest range for which it was designed?
16. If readings obtained on the vtvm are erratic, what might be the cause?

CHAPTER 8

Solid-State Analog VOMs

The solid-state analog volt-ohm-milliammeter is sometimes referred to as a transistorized vtvm. Solid-state analog instruments perform much like the vtvm; they offer most of the advantages of both the vom and the vtvm. They are lightweight, compact, battery-operated, portable, versatile, and they require no warmup time. The active components in solid-state instruments are either conventional bipolar transistors, field-effect transistors (FETs), or a combination of both. Solid-state analog instruments are variously referred to as transistor voltmeters (tvm's); transistor volt-ohm-milliameters (tvom's), electronic multimeters (emm's), field-effect transistor vom's (FET-vom's), solid-state vom's (ssvom's), and so on. In most cases we will refer to all of these as (transistor volt-ohm-milliameters).

SPECIAL FEATURES OF SOLID-STATE INSTRUMENTS

Most tvom's have an 11-megohm input impedance, the same as the typical vtvm; however, some have a higher input impedance, such as 15 megohms or 21 megohms. As compared to the vtvm, the typical tvom can measure lower dc voltages, is more stable, offers greater portability, can provide "low-power" resistance measurement, and seldom needs to be reset to electrical zero. The tvom has replaced the vtvm to a great extent for many uses. Also, except where low cost is important, the tvom is slowly replacing the conventional vom.

CIRCUITS OF TYPICAL INSTRUMENTS

Some typical tvom's will be examined in this part of the book. Since most tvom's utilize FETs, let's stop to examine briefly how the FET operates.

The operating principle of the FET, schematically shown in Fig. 8-1, is as follows. A negative voltage, which is similar in action to the bias of a vacuum tube, is applied between the *gate* and the *source* of the FET. Also, a positive voltage, similar to the plate voltage of a vacuum tube, is applied between the *drain* and the source, establishing a current between source and drain. When the gate is

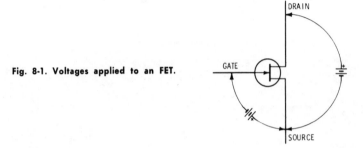

Fig. 8-1. Voltages applied to an FET.

biased negatively enough, "pinch-off" (like "cutoff" in a tube) occurs, stopping the drain current. With typical operating voltages applied to the gate, source, and drain, a more-negative gate bias results in less drain current, and a less-negative gate bias results in more drain current. The negative voltage between gate and source results in negligible gate current; thus, the gate-to-source impedance is high. The input signal is normally connected between the gate and the source. It is its high input impedance that makes the FET popular in solid-state measuring instruments.

A simplified solid-state metering circuit is shown in Fig. 8-2. Transistor Q1 is an FET which provides a high-impedance input for the dc voltage to be measured. Transistors Q2 and Q3 are part of the amplifier that drives the meter, M1.

For the measurement of ac voltages, the same basic circuit may be employed, but with the addition of a rectifier circuit as for the vom and vtvm. For lowest-range ac measurement, sometimes an additional amplifier stage is added.

A solid-state metering circuit also can be based on the arrangement of Fig. 8-2. In most respects it is similar to the ohmmeter circuits of the vom and vtvm.

Fig. 8-3 shows an example of a FET-tvom, the EICO Model 242. It provides for measurement of dc voltages between 0.01 and 1000

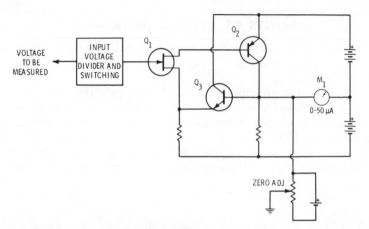

Fig. 8-2. Simplified solid-state metering circuit.

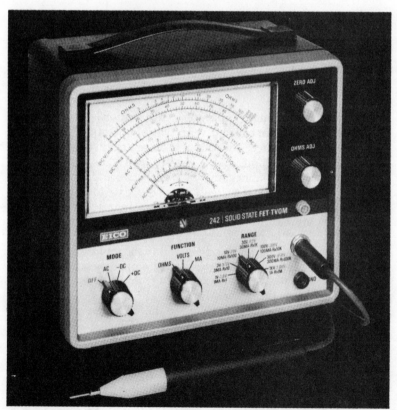

Courtesy EICO, Electronic Instrument Co.

Fig. 8-3. EICO Model 242 FET-tvom.

volts and up to 30,000 volts with the addition of a high-voltage probe. Voltage ranges for ac cover from 1 volt rms full scale to 1000 volts rms full scale and from 25 Hz to 2 MHz or up to 250 MHz with the use of an rf probe. A low-voltage source permits measurement of resistance in transistor circuits without introducing error due to transistor conduction, and it lessens the chance of damage to the transistors being tested. Choice of operation either from self-contained batteries or ac power is provided.

The input circuit of the instruments includes a FET and a differential amplifier circuit, as shown in simplified form in Fig. 8-4. The FET, connected as a source-follower stage (Q1), drives a differential amplifier consisting of silicon transistors Q2 and Q3. Whether the parameter being measured is current, resistance, or voltage, the meter reading on M1 is a function of the dc voltage applied to the Q1 FET. When the emitter currents through Q2 and Q3 develop identical voltages across their respective emitter resistors, R18 and R19, the two emitter voltages will be identical. No current will flow through the meter at this time. Note that the base current of Q3 is held constant by voltage-divider elements R20 and R21, thus maintaining the emitter current and the voltage developed across R19 at a fixed value.

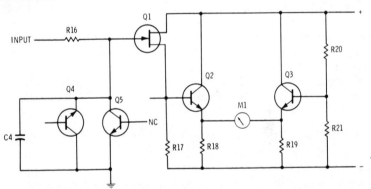

Fig. 8-4. Basic FET-tvom circuit for the EICO Model 242.

The operation of Q2 is controlled by a voltage divider consisting of FET Q1 (acting as a voltage-variable resistor) and R17. When a dc voltage is applied to the gate of source follower Q1, its source voltage, direct coupled to the base of Q2, changes accordingly. Since Q2 is connected as an emitter follower, its emitter voltage follows the change. This unbalances the voltages between the emitters of Q2 and Q3 and deflects the meter pointer. In this manner, meter deflection is made a function of the dc voltage applied to the gate of Q1.

111

Transistors Q4 and Q5 are used in conjunction with resistor R16 to protect FET Q1 from accidental overloads. With the bases of Q4 and Q5 disconnected, they act as high-quality temperature-compensated zener diodes. This parallel transistor circuit presents an infinite impedance to input voltages up to approximately 10 volts, but it becomes a short circuit to higher voltages. Excess voltages are dropped across R16.

The voltage-measuring circuit of the FET-tvom is generally similar to that of the typical vtvm, so we will not go into voltage-measuring circuits here. Many FET-tvom's and other solid-state instruments provide for measurement of direct current and sometimes alternating current; the instrument being considered here provides for measurement of both ac and dc current. An analysis of these circuits is worthwhile at this time.

Input Circuit for Measurement of DC Milliamperes

Direct-current measurements ranging from 0.01 milliampere (10 microamperes) to 1 ampere can be made with the Model 242. These measurements are made by passing the current through a resistance of known value, then measuring the voltage drop across the resistance.

Fig. 8-5 shows the input circuit arranged to measure up to 1 milliampere of current. When the test probe and common lead are connected in series with the circuit to be tested, the external current flows through resistors R31 through R34. Assuming that 1 milliampere of current flows, a voltage drop of 1 volt (1 milliampere × 1000 ohms) is produced. The 1-volt dc level is then applied through R35 and voltage-divider elements R8 through R13, R7, R6, R5, and R22. This 11-megohm network is identical to that used for measuring dc voltage, except that R35 replaces probe resistor R1. With $\frac{10}{11}$ volt fed through R13 to FET Q1, full-scale deflection is produced on the 1-mA dc scale.

When RANGE switch S1 is set to the 3-mA position, current flows through the same network (R31 through R34). Assuming that 3 mA of current flows in the external circuit, 3 volts dc is developed across this network. The 3-V tap on the voltage divider is automatically selected at this time.

When measuring 10 mA or 30 mA, the external current flows through R31, R32, and R33, again producing 1 volt or 3 volts, respectively. Again, the same two taps are used on the voltage divider to produce the voltage for proper deflection.

Operation is similar on the remaining dc current scales. In each case, either 1 volt or 3 volts is developed and either of the same two voltage-divider taps is selected.

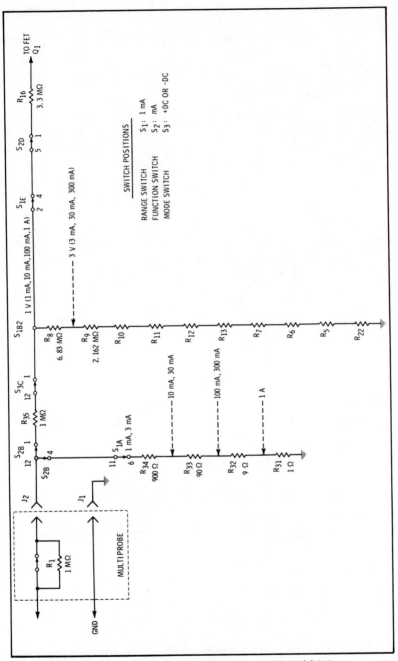

Fig. 8-5. Dc milliampere input circuit for the EICO Model 242.

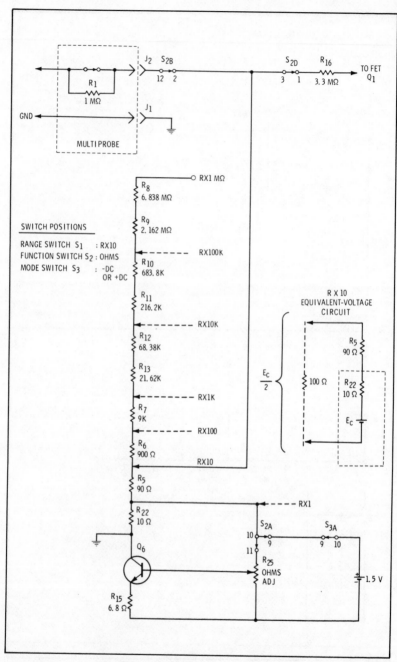

Fig. 8-6. Ohms input circuit of the EICO 242.

Circuit for Measurement of AC Milliamperes

The ac milliampere circuit functions similarly to its dc counterpart. The same resistors (R31 through R34) are used to develop either 1 volt or 3 volts ac. In this case, the ac voltage is coupled through a blocking capacitor to a voltage-doubler circuit where it is rectified to dc. The dc voltage is then applied to divider resistors as before, and the proper voltage is tapped off at either the 1-V or the 3-V tap for proper deflection on the range in use.

FET-TVOM Ohmmeter Circuit

As shown in Fig. 8-6, the ohmmeter circuit of the EICO Model 242 uses a constant-current source consisting of transistor Q6, OHMS ADJ control R25, resistor R15, and the 1.5-volt battery; all of these acting together maintain the current through collector load resistor R22 at a constant value. The bias for the base of Q6 is set by means of the OHMS ADJ control so that, with no external resistance connected, 0.91 volt is applied to the FET Q1 gate, producing full-scale deflection, designated as infinity (∞) on the meter face. However, with the ohmmeter in the R × 10 position, as shown, if an external 100-ohm resistor is connected between the probe and the common test lead, the voltage at the gate of Q1 is reduced to one-half its original value. This is shown in the equivalent voltage circuit of Fig. 8-6. Therefore, measurement of 100 ohms on the R × 10 scale causes the meter to point to the halfway mark, or 10, on the OHMS scale. If the external resistance is lower in value, less voltage is fed to the gate and a lower ohms reading is obtained.

The resistance connected in series with the constant-voltage source increases by a factor of 10 as the RANGE switch is rotated through each position, from R × 1 to R × 1 M. Resistance measurement above 1000 megohms can be made by employing an external battery or voltage source between 20 and 500 volts, using the same procedure as described earlier for extending the resistance-measuring range of the vtvm.

The B&K Model 277 Solid-State Electronic Multimeter

Another solid-state electronic multimeter employing a FET input circuit is the B & K Model 277, shown in Fig. 8-7. One of the special features of this instrument is that it provides for measurement of dc current down to 0.1-microampere full scale. Also included are high- and low-power ohms ranges for critical testing of semiconductor circuits.

The B & K Model 277 also includes conventional high-power resistance-measuring ranges since these are necessary when trying to determine if the transistor in the same circuit is good or bad. With the high-power resistance-measuring circuit in use and the test leads connected first one way across the transistor terminals, then reversed, the two readings are compared to obtain the front-to-back ratio. This ratio provides a good indication of the condition of the transistor.

The RCA WV-500B Solid-State VoltOhmyst

The solid-state RCA WV-500B VoltOhmyst, shown in Fig. 8-8, does not utilize FETs, but rather employs a four bipolar-transistor amplifier circuit designed especially for good linearity and stability. The manufacturer's description of the circuit (shown in Fig. 8-9) is based on the following descriptions.

Metering Circuit—The input voltage (from the ac/dc voltage divider or ohms divider) is applied across the bases of Q3 and Q4, positive to Q3 base negative to Q4 base. These transistors provide a nearly infinite impedance. This high impedance is achieved through

Courtesy of B & K Precision Division of Dynascan Corp.

Fig. 8-7. B & K Precision Model 277 solid-state electronic multimeter.

a controlled positive feedback network, R13A, B, R32A, B, and the two impedance-adjust potentiometers, R45 and R46. Transistors Q3 and Q4 serve as preamplifiers driving the bases of Q1 and Q2. In effect, transistors Q3 and Q1 amplify the positive portion of the signal, and Q4 and Q2 amplify the negative portion of the signal.

Negative feedback through R28 and R29 results in high impedance at the Q1 and Q2 bases to prevent loading the Q3 and Q4 emitters. The outputs of Q1 and Q2 drive the 50-microampere meter.

Potentiometer R30 is used to balance the Q3/Q4 input. This is a factory adjustment and need not be readjusted unless the transistors are replaced. The front-panel zero control, R21, balances the amplifier output with no input signal applied.

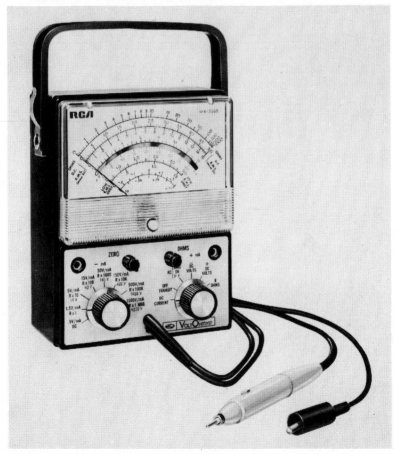

Courtesy RCA

Fig. 8-8. RCA Model WV-500B solid-state VoltOhmyst.

Resistors R33 and R34 serve to isolate and protect the amplifier circuit. Capacitors C5 through C10 are bypass capacitors to prevent ac signals from affecting the metering circuit.

The accuracy of the WV-500B is maintained throughout the usable life of the batteries. Since current drain of the instrument is very low, battery life is approximately equivalent to shelf life; there is no provision for powering the instrument from the ac line.

Silicon diodes CR3 and CR4 are connected across the meter terminals to prevent meter damage due to accidental overload.

DC Voltage Circuit—The dc voltage input is applied through the isolation resistor in the probe to the voltage-divider network (range), resistors R11 through R17. The voltage from the divider network is then connected to the transistorized metering circuit.

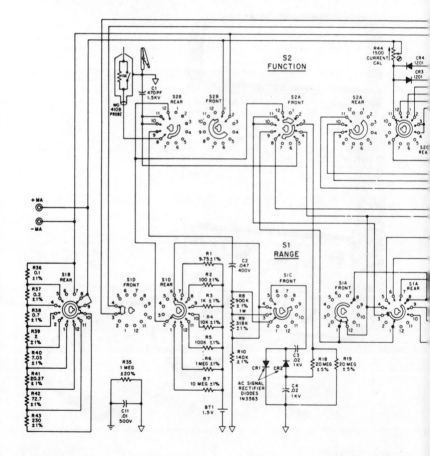

Fig. 8-9. WV-500B

DC Current Circuit—The transistor amplifier is not used in the dc current function. A separate shunt resistor (R36 through R43) is connected across the meter for each of the current ranges. The current input is connected directly to the meter and shunt circuit. Potentiometer R44 is used to calibrate the current-measuring function.

AC Voltage Circuit—When the VoltOhmyst is used to measure ac voltage, the signal is first rectified by diodes CR1 and CR2 which form a full-wave peak-to-peak rectifier. The circuit components are chosen to provide a long time constant. When the signal swings negative, C3 is charged through CR1 to the negative peak value of the voltage. As the input signal starts in a positive direction, C4 charges to a value equal to the sum of the positive and negative peaks. Because of the relative time constant, the voltage across C4

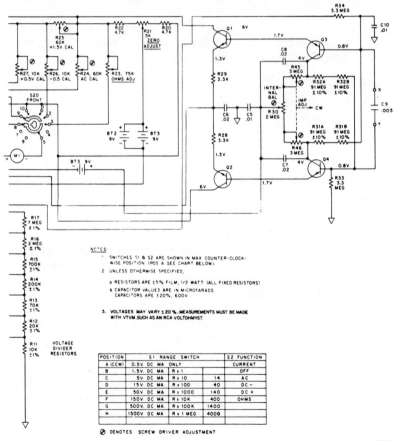

NOTES:

1. SWITCHES S1 & S2 ARE SHOWN IN MAX COUNTER-CLOCK-WISE POSITION (POS A SEE CHART BELOW)

2. UNLESS OTHERWISE SPECIFIED,

 a. RESISTORS ARE ±5% FILM, 1/2 WATT (ALL FIXED RESISTORS)

 b. CAPACITOR VALUES ARE IN MICROFARADS
 CAPACITORS ARE ±20%, 600V

3. VOLTAGES MAY VARY ±20 %. MEASUREMENTS MUST BE MADE WITH VTVM. SUCH AS AN RCA VOLTOHMYST.

POSITION	S1 RANGE SWITCH			S2 FUNCTION
A (CCW)	0.5V. DC MA ONLY			CURRENT
B	1.5V. DC MA	R x 1		OFF
C	5V. DC MA	R x 10	14	AC
D	15V. DC MA	R x 100	40	DC −
E	50V. DC MA	R x 1000	140	DC +
F	150V. DC MA	R x 10K	400	OHMS
G	500V. DC MA	R x 100K	1400	
H	1500V. DC MA	R x 1 MEG	4000	

⌀ DENOTES SCREW DRIVER ADJUSTMENT

Courtesy RCA

schematic diagram.

will be maintained at the peak-to-peak value of the ac signal. This signal, now, dc is fed through the voltage-divider network and then to the metering circuit.

Resistance Circuit—The voltage from battery BT1, 1.5 volts, is applied through the selected ohms divider resistor (R1 through R7) to the external resistance under test. A voltage divider is formed by the range resistor and the external resistance. The output of this divider is fed to the metering circuit.

Simpson Model 313-2 Solid-State VOM

The Simpson Model 313-2 solid-state vom is shown in Fig. 8-10. It includes a larger, 7½-inch meter, a mechanical "on" indicator to show when the instrument is powered without contributing to battery drain, a BATT TEST position on the function-selector switch to show the condition of the 9-volt battery, an input resistance of 11 megohms, an overload protection, a multifunction test probe (not shown), and a shielded cable which plugs into the type BNC connector at the lower center of the instrument.

Triplet Model 603 VOM

The Triplett Model 603 vom is shown in Fig. 8-11. According to the manufacturer, this instrument can be left on for as long as one year or more and still operate satisfactorily from the same 9-volt battery. This model includes a low-power ohms circuit and an automatic-polarity circuit which allows the user to make dc measurements without having to worry about polarity, thus saving time in many cases where a number of measurements are to be made. For conventional polarity measurements, either the + or the − push button at the lower-left side of the front panel is depressed. For automatic polarity, both of these are depressed at the same time. Two types of batteries are used in this instrument, a D cell for resistance measurements and two 9-volt batteries for the amplifier circuit. When the range switch is in the BATT CHECK position, the condition of one of the 9-volt batteries may be observed on the meter dial when the + push button is depressed; the condition of the other 9-volt battery is observed when the − push button is depressed. If the pointer falls within the BATT OK range shown below the upper end of the scale, the battery being tested at that time is in satisfactory condition. The best test for the condition of the D cell is to see whether the pointer can be brought to zero ohms with the instrument in any of the resistance-measuring positions. If the pointer cannot be brought to zero by adjustment of the OHMS ADJ control, the D cell should be replaced. An additional check of the D cell is recommended by occasionally measuring a resistor of known value between 5 and 10 ohms on the × 1 range. If the resistance reading is

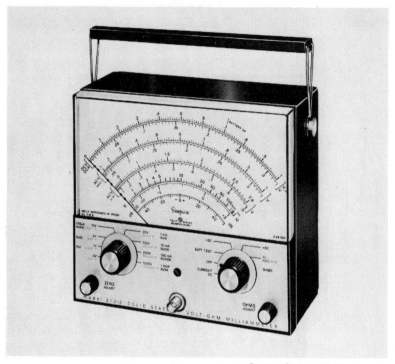

Courtesy Simpson Electric Company

Fig. 8-10. Simpson Model 313-2 solid-state vom.

low, the D cell should be replaced because the low reading indicates that the internal resistance of the battery is too high causing the low reading.

Sencore FE27 Field-Effect Multimeter

One manufacturer provides a series of multimeters that utilize a number of push-button switches rather than one or more rotary switches. An example is the Sencore Big Henry FE27 Field-Effect Multimeter, shown in Fig. 8-12. The FE27 has a 15-megohm input resistance and has been designed to withstand unusual degrees of both mechanical shock and electrical overload. Accuracy is 1.5% on dc voltages, 3.0% on ac voltages, and 2% of arc on ohms measurements. The lowest dc voltage range is 0.3-volt full scale.

CARE OF SOLID-STATE INSTRUMENTS

Solid-state instruments require the same careful handling as the vom and vtvm, since they contain the same type of meter movement.

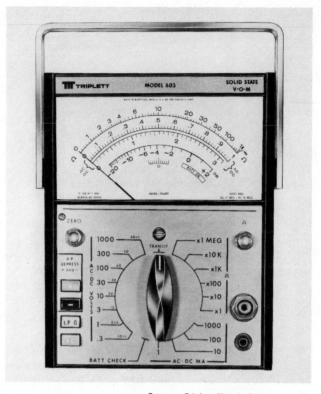

Courtesy Triplett Electrical Instrument Co.

Fig. 8-11. Triplett Model 603 solid-state vom.

Also, they are affected by high extremes of temperature and humidity.

Batteries in solid-state instruments can be expected to last six months to one year or more. When the instrument is not in use, it usually should be turned off to extend battery life. Batteries may have their useful life shortened somewhat if they are operated in abnormally warm surroundings. If operated in cool or cold surroundings, batteries will last longer. But at about 0°F and below, battery capacity will decrease; below about −20°F dry-cell batteries will provide no output.

When batteries are replaced, care should be used to be sure that proper polarity is observed. In some instruments, that batteries are soldered into the circuit to ensure proper and continuous contact. When soldering leads to batteries, first "tin" the battery terminals; that is, apply a small coating of solder to the positive and negative terminals. Then it will be easier to solder the leads to the terminals.

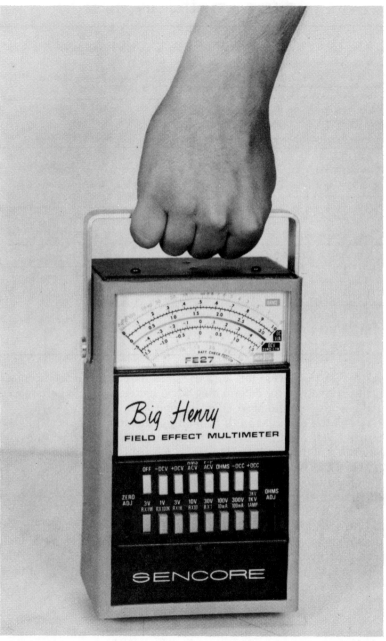

Courtesy Sencore, Inc.

Fig. 8-12. Sencore Model FE27 field-effect multimeter.

Table 8-1. Troubleshooting Hints

Instrument fails to operate on all functions. 9-volt batteries dead. Probe defective. Check continuity.	Instrument fails to indicate properly on any ac voltage range; operates normally on dc volts and ohms. Diodes CR3 or CR4 defective. C1, C10, C11, R35 to R37 defective.
Voltage readings low on battery operation, especially at right-hand side of meter scale. 9-volt batteries require replacement.	Instrument inaccurate on 500- and 1500-volt ac ranges. Resistors R35, R36, or R37 defective.
Meter cannot be adjusted full-scale on ohms function. 1.5-volt battery weak.	Instrument inaccurate on any ac or dc voltage range. Ohms all right. Check voltage divider resistors, R40 through R46.
Instrument cannot be zeroed on any voltage function. Check transistor amplifier circuit.	Resistance readings inaccurate. Check resistors R28 through R34. Poor connection to 1.5-volt battery. High internal resistance in battery.
Meter pointer bangs hard left or right, depending on function switch setting. Open or short circuit in amplifier metering circuit.	Instrument inaccurate or inoperative on current ranges only. Check resistors R19 through R26.
Meter pointer sticks, or is sluggish on all functions. Meter movement defective.	

Instruction manuals for various solid-state instruments contain further directions for determining when batteries should be replaced and how they should be installed or connected.

Check the manufacturer's manual for directions on removing the instrument from the case, for changing fuses, for calibrating and adjusting, and for setting the pointer mechanical zero adjustment.

When the tvom is not in use, set the function-selector switch to the off position (if there is one) to conserve battery life and also to shunt the meter to prevent it from moving excessively while in transit.

Instruction manuals also include troubleshooting hints for determining whether the batteries or other components are at fault when problems arise. Typical of such troubleshooting hints is the information in Table 8-1, which is provided by RCA for users of their Master VoltOhmyst Model WV-510A.

QUESTIONS

1. Name some of the advantages of solid-state voltmeters and vom's.
2. What is a FET?

3. What feature of the FET makes it popular for use in instruments used for measuring?
4. Describe the purpose of the gate, source, and drain in the FET.
5. Sketch a schematic of a simple solid-state metering circuit.
6. If the function switch includes a BATT position, what is the purpose of this position?
7. What is the input impedance of a typical solid-state vom?
8. What general function does the FET serve in an electronic vom?
9. How long do batteries generally last in solid-state instruments?
10. What is the lowest temperature at which battery-powered instruments can generally be used?

CHAPTER 9

Digital VOMs

A digital vom displays a voltage, a current, or a resistance value, not as a quantity indicated by a pointer along a calibrated scale but as an exact or a discrete value. In one example of a digital vom, shown in Fig. 9-1, there is no judgment factor involved (as there would be with an analog instrument) in deciding that the value displayed is 18.52.

Note, also, that the + sign to the left of the 18.52 indicates that the value is positive. Furthermore, since the FUNCTION switch is set on DC OHMS, and the RANGE switch is set on 20 mA, the value is more specifically understood to be 18.52 mA, dc.

Another example of a digital vom, or dvm, is the MITS Model DV1600 in Fig. 9-2. It is shown displaying a value of 19.6 volts dc.

Both of the dvm's shown include an automatic-polarity feature, meaning that it is not necessary to be concerned whether or not the common test lead is connected to the negative side and the other test lead to the positive side of the voltage or current source to be measured. The display includes a polarity sign, to indicate whether the "hot" test lead is connected to the negative ($-$) or positive ($+$) side of a circuit. Not all dvm's include this automatic-polarity feature; on some of those that do not, a polarity-reversing switch is provided.

BASIC COMPONENTS OF DIGITAL VOMS

As shown in Fig. 9-3, the basic components of the digital vom are a signal-conditioner stage, an analog-to-digital converter (adc) stage, and a readout or display stage. Each of these stages includes other features which will be discussed later.

Signal Conditioner

At the input, the first stage, which is the signal conditioner, includes a function switch that is set by the user for the type of measurement desired, −dc, +dc, ac, current, or ohms. Also included in the signal-conditioner circuit is a range switch having the same purpose as the range switch in an analog instrument; that is, to reduce the input signal, if necessary, to a value that is within the measurement capability of the next stage (the analog-to-digital converter). (Typically, the adc has an input limitation of 1 volt.)

Courtesy Simpson Electric Company

Fig. 9-1. Simpson Model 360 digital vom.

Some dvm's include "range-changing" circuits. The signal conditioner may also amplify a signal if its level is below that signal level desired. In the signal conditioner, all inputs, whether negative dc voltage, ac voltage, current, or resistance, are converted to positive dc voltages before reaching the adc. Thus, basically, the dvm reads only positive dc voltages.

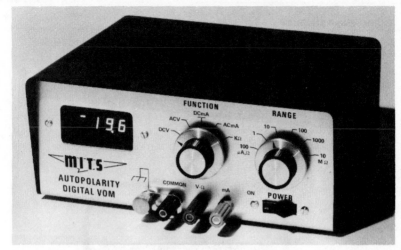

Courtesy MITS, Inc.

Fig. 9-2. MITS Model DV1600 digital vom.

Analog-to-Digital Converter

In the adc, the dc signal being measured is changed from one that may vary continuously over a 0- to 1-volt range to a voltage that varies only in steps or in discrete amounts. Some of the different types of adc's will be discussed later.

Display or Readout

The display or readout is the last basic stage of the dvm. The readout consists of either vacuum-tube, gaseous, or solid-state devices which display a series of self-illuminating numbers between 0 and 9 to show the numerical value of whatever quantity is being

Fig. 9-3. Basic components of a digital vom.

measured. The display device is most often a light-emitting diode (LED), but other types, such as liquid-crystal, incandescent, gas-discharge, or Nixie displays, are also used. A minus sign, a plus sign, and/or a decimal point may also be displayed.

Individual numbers in the display are made up of straight-line segments. This means that any number can be produced by lighting the right combination of segments. The arrangement of segments in

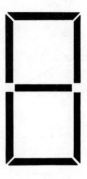

Fig. 9-4. Arrangement of a 7-segment display as used in a digital voltmeter readout.

a 7-segment display is shown in Fig. 9-4. The manner in which different segments are illuminated to produce numbers 0 through 9 is shown in Fig. 9-5.

Fig. 9-5. How different segments of a 7-segment display are illuminated to produce numbers 0 through 9.

TYPES OF ANALOG-TO-DIGITAL CONVERTERS

There are a number of adc circuits, including single linear ramp, dual-slope integration, staircase ramp, and intergrating. In all of these, an analog signal is converted to a digital signal. As a simplified example, an analog voltage value is changed into a corresponding interval of time which is used to start and stop an accurately controlled oscillator. The output pulses from the oscillator are fed to a digital counter. The counter has a display that indicates a digital value that is equal to the value of the analog voltage being measured.

The concept of the linear-ramp type of adc is shown in Fig. 9-6. In this example, the voltage ramp extends from +2 V through 0 to −2 V during an accurately controlled period of time. The voltage to be measured is shown to be 1.8 volts. When the value of the ramp voltage reaches 1.8 volts—that is, when the ramp voltage and the voltage to be measured are equal—a gating pulse is started and an oscillator or clock pulse generator is turned on. The oscillator is turned off when the ramp reaches zero, which is when the gating pulse is turned off. The number of pulses, in this case 10, generated during the gating-pulse time interval, are in direct relation to the voltage to be measured. The pulses are fed to a digital counter readout, which in this example would indicate 1.8 volts.

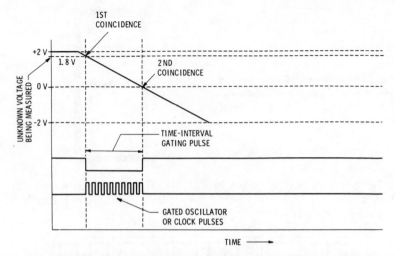

Fig. 9-6. Principle of single-ramp method of analog-to-digital conversion.

The dual-slope, or dual-ramp, adc is perhaps the most widely used. It is a little more complex but somewhat more accurate than the single-ramp method which can be adversely affected by noise.

The principle of the dual-slope, or double-ramp, method of adc, shown in Fig. 9-7, is adapted from the principles of operation of the Philips Model PM2421. The signal to be measured is applied to a capacitor in the input circuit of the instrument. The capacitor is

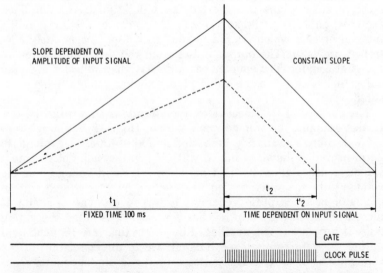

Fig. 9-7. Principle of dual-slope integration method of analog-to-digital conversion.

130

charged by a current proportional to the input signal for a period of 100 milliseconds (ms). The resulting charge is proportional to the mean value of the signal during this interval. At the end of t_1, the charge is then dissipated at a constant-current rate (indicated by "constant slope" in Fig. 9-7) for time t_2, which will also be proportional to the input signal. Over this period t_2, a gate is actuated, providing a series of pulses which are counted and indicated on a digital readout. The readout figure shows the value of the input signal.

In the staircase ramp adc, a sample of the input voltage is compared to an internally generated "staircase ramp" voltage. A block diagram of the Hewlett-Packard Model 3430A digital voltmeter, an instrument employing this method, is shown in Fig. 9-8.

When the input signal, or voltage to be measured, and the staircase ramp voltages are of the same value, the comparator generates a signal to stop the ramp. The instrument readout then displays the number of counts necessary to make the staircase ramp equal to the input voltage. At the end of the sample, a reset pulse from the 2-Hz sample oscillator resets the staircase to zero and the measurement is repeated starting with a new sample. The display shows the value of the previous sample until the next one is completed; the process is repeated every half second.

INTEGRATING TYPE OF ANALOG-TO-DIGITAL CONVERTER

In the integrating type of adc, the average of the input voltage is measured over a predetermined measuring period. By use of an integrating circuit at the input, the average value of the voltage to be measured is attained for the measuring interval. Integration allows for relatively accurate measurement of the input voltage even with substantial interference from superimposed noise. The main advantage of this type of converter is its relative immunity to noise interference.

The integrating circuit normally includes an amplifier, designated as $-A$ in Fig. 9-9, as part of the feedback control system in a voltage-to-frequency converter. The feedback control system governs the pulse repetition rate of the pulse generator so that the average voltage of the generated train of rectangular pulses is equal to the dc input voltage. As shown, the level detector regulates the rate or frequency of the pulse generator.

RESISTANCE MEASUREMENT

In the digital vom, the resistance-measurement circuit includes a source of constant current which flows through the unknown resis-

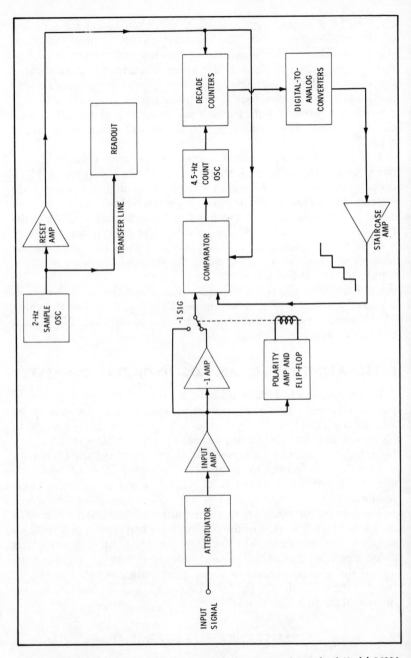

Fig. 9-8. Staircase-ramp method of conversion used in the Hewlett-Packard Model 3430A digital voltmeter.

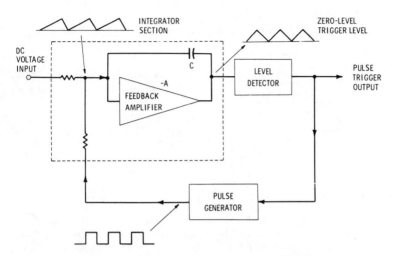

Fig. 9-9. Voltage-to-frequency converter in integrating-type analog-to-digital converter.

tor; then the internal circuitry measures the voltage drop across the resistor and provides a readout of the value of the resistance in ohms.

A particular advantage of the digital vom is its ability to measure resistance, especially low values, much more accurately than the average analog vom.

DIGITAL VOLTMETER SPECIFICATIONS

Digital vom specifications include some that do not apply to, or cannot be interpreted the same way as, those associated with the analog vom's. These specifications are the accuracy, the resolution, the common-mode rejection, and the normal-mode rejection.

Accuracy

The accuracy of a digital vom is usually stated in terms of a percentage-of-reading error. There is an additional possible error in the reading of plus-or-minus one digit. For example, a certain digital vom may have accuracy specified as ±0.75% of full-scale reading, ±1 digit. Thus, for a reading of 130.5 volts on a meter having a 4-digit display, the 0.75% rating would mean that the actual value might be off as much as $130.5 \times 0.0075 = 0.98$ volt, or approximately 1.0 volt in either the plus or the minus direction. The plus-or-minus-one-digit rating means that the 130.5-volt value could be anywhere between 130.45 and 130.55.

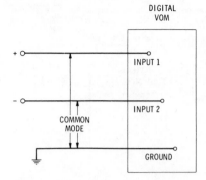

DIGITAL
VOM

+ ○

INPUT 1

– ○

INPUT 2

COMMON
MODE

GROUND

Fig. 9-10. How common-mode interfering signals can enter the input circuit of a digital vom.

Resolution

The resolution rating of a digital vom refers to the dynamic range of a vom. In a 5-digit vom, the resolution can be 1 part in 100,000 and may be specified as 0.001%. In a 4-digit vom, the resolution is 1 part in 10,000, or 0.01%.

Common-Mode and Normal-Mode Rejection

In digital vom's having both input terminals isolated from the case of the instrument, an interfering or unwanted signal (say, 60 Hz) could enter the input circuit through both input terminals with respect to ground, as shown in Fig. 9-10. The ability of a digital vom to reject such common-mode signals is referred to as its common-mode rejection rating and is often specified in decibels.

The ability of a digital vom to prevent line-frequency interference, noise, and other stray signals from entering the input circuit along with the desired signal is referred to as normal-mode rejection.

DIGITAL VOM RANGES AND "OVERRANGING"

When we change ranges in the analog vom, we either switch to another scale or use another multiplier, or both. With a digital vom, there are obviously no scales. Instead, we must observe the location of the decimal point which moves with each different range. We must also be familiar with other digital vom characteristics referred to as half-digit capability and "overranging." For instance, an instrument having 5 digits in the readout display is referred to as a 4 ½-digit instrument; one having 4 digits is a 3½-digit instrument, and so on. This is because all of the digits except the most significant digit (the one on the left) can have any value between 0 and 9. These are referred to as the least significant digits. The most significant digit can be only 0 or 1; therefore, it is referred to as a "half digit."

The overrange specification for a digital vom is usually somewhere between 10% and 100%. Overrange may be understood from the following example. In a certain instrument, on the 1000-V setting of the range switch, the left-most digit can go only as high as 1, but all the others can go as high as 9. Therefore, if the electronic circuitry has been provided in the instrument, the readout can be anything up to 1999 V, or virtually 100% beyond the 1000-V setting of the range switch. Many instruments are provided with 100% overrange capability, but some with 10% only. In the preceding example, a 10% overrange specification would allow an upper-limit reading of only 1100 volts.

If there is an attempt to make a measurement higher than what the instrument is capable of reading at a particular range setting, the usual digital vom will provide some means to warn the user of the overload. In some instruments, the most significant digit will flash on and off continually; in others a separate overrange warning light comes on.

TYPICAL INSTRUMENTS AND CIRCUITS

In the following pages we will consider a number of typical digital voms and discuss their features and circuits. In some cases the features and circuits will be examples of those previously discussed. In other instances, we will be seeing or studying new or improved design features.

Leader Model LDM-850 Digital Multimeter

Fig. 9-11 is a front view of the Leader Model LDM-850 digital multimeter, a 3½-digit instrument. Views of the top and bottom, with covers removed, are shown in Figs. 9-12 and 9-13, respectively. The front view includes identifying the controls and their purposes. The back of the instrument is provided with legs for vertical operation. Also, on the back panel is a fuse for protection against current or voltage overloads in current- and resistance-measuring circuits. The ac line cord also exits from the rear panel.

Display Switches

The DISPLAY HOLD switch at the lower right in Fig. 9-11 is normally off, but it may be switched on at any time to hold whatever value is being displayed. The value will then continue to be displayed until the switch is returned to the off position. When the DISPLAY CHECK switch, also at the lower right, is depressed, the display reads +1 8 8 8, which causes all segments in the display tubes to light up. This provides an immediate and continuously available check on the operation of the display section. The decimal point can be checked

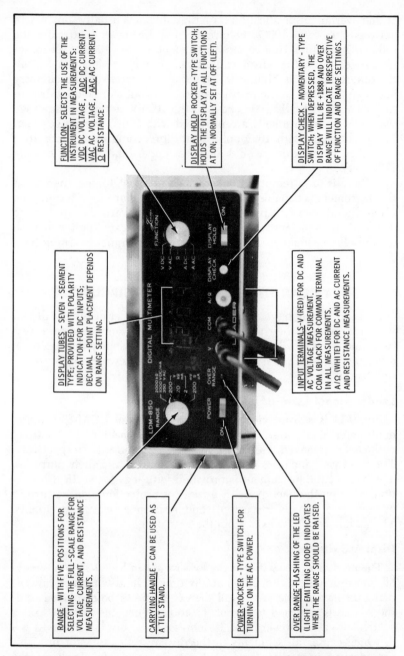

FUNCTION- SELECTS THE USE OF THE INSTRUMENT IN MEASUREMENTS: VDC DC VOLTAGE, ADC DC CURRENT, VAC AC VOLTAGE, AAC AC CURRENT, Ω RESISTANCE.

DISPLAY HOLD- ROCKER-TYPE SWITCH; HOLDS THE DISPLAY AT ALL FUNCTIONS AT ON; NORMALLY SET AT OFF (LEFT).

DISPLAY CHECK - MOMENTARY - TYPE SWITCH; WHEN DEPRESSED, THE DISPLAY WILL BE +1888 AND OVER RANGE WILL INDICATE IRRESPECTIVE OF FUNCTION AND RANGE SETTINGS.

DISPLAY TUBES - SEVEN - SEGMENT TYPE; PROVIDED WITH POLARITY INDICATION FOR DC INPUTS; DECIMAL - POINT PLACEMENT DEPENDS ON RANGE SETTING.

INPUT TERMINALS -V (RED) FOR DC AND AC VOLTAGE MEASUREMENT, COM (BLACK) FOR COMMON TERMINAL IN ALL MEASUREMENTS. A/Ω (WHITE) FOR DC AND AC CURRENT AND RESISTANCE MEASUREMENTS.

RANGE - WITH FIVE POSITIONS FOR SELECTING THE FULL-SCALE RANGE FOR VOLTAGE, CURRENT, AND RESISTANCE MEASUREMENTS.

CARRYING HANDLE - CAN BE USED AS A TILT STAND.

POWER- ROCKER - TYPE SWITCH FOR TURNING ON THE AC POWER.

OVER RANGE-FLASHING OF THE LED (LIGHT - EMITTING DIODE) INDICATES WHEN THE RANGE SHOULD BE RAISED.

Courtesy Leader Instruments

Fig. 9-11. Front-panel operating controls of the Leader LDM-850 digital multimeter.

Courtesy Leader Instruments
Fig. 9-12. Top view of LDM-850 with cover removed.

at each position by moving the Range switch, upper left in Fig. 9-11, through each setting.

The OVERRANGE lamp, lower left of center in Fig. 9-11, will light when the value being measured is too high for the range selected; therefore, a higher range must be used. This instrument also includes automatic polarity determination; either a plus or a minus precedes voltage and current readings depending on which side of the circuit being measured is contacted by the red test lead. Also included is an overload protection and an automatic decimal-point placement.

Measurement Capabilities

The LDM-850 can measure ac and dc voltage and current as given in Chart 9-1.

Courtesy Leader Instruments

Fig. 9-13. Bottom view of LDM-850 with cover removed.

Resistance Measurements Using the LDM-850

The following directions for measurement of resistance are taken from the operating manual for the LDM-850. They are fairly typical of those given for most other digital vom's.

Resistance Measurements—Maximum Resistance: 1999kΩ (1.999MΩ).

1. Connect the black test lead to COM terminal and the red test lead to A/Ω (white) terminal.
2. Set FUNCTION switch at Ω.
   When the test leads are open, there will be an overrange indication.
3. Set RANGE switch at the range under measurement. If the range is not known, set at 2000kΩ and lower the range as required.

Chart 9-1. LDM-850 Measurement Capabilities

Function	Measurement Capabilities
Dc Voltage	± 0.1 mV to ± 1000 V
Ac Voltage	0.1 mV to 350 V rms
Resistance	0.1 Ω to 1999K (1.999 MΩ)
Dc Current	± 0.1 μA to ± 1000 mA
Ac Current	0.1 μA to 1000 mA

4. Connect the test lead prods across the resistor or device under test. If the resistor is wired in a circuit, turn off the power before measurement.
5. Read the resistance, kΩ, or Ω, referring to the RANGE switch setting.

NOTES: 1. When overrange is indicated during the measurement, lower the range.
2. The figure 00.1, 00.2, or 00.3 will be displayed when the test leads are shorted at the 200 Ω range. This represents the resistance of the test leads and terminal contact; it is a normal condition. The value should be subtracted from the displayed value, especially when low resistances are being measured.
3. When measuring resistors wired in a circuit, make certain that there are no other resistors or semiconductors in parallel. Check the schematic.
4. Overload protection (FUNCTION at Ω):

    A. 200 Ω range. The protective fuse will blow if, in error, more than 1 A rms or 1 V rms is applied across the input.
    B. Other ranges. Will withstand application up to 100 V rms across the input.

Diode and Transistor Checking—Tests for "quality," or forward/backward resistance characteristic, can be made on diodes and transistors under low current conditions. The procedure is the same as given for resistance measurements.

The maximum currents applicable depend on the RANGE switch settings as given in Chart 9-2.

NOTES: 1. The A/Ω terminal is the plus (+) side.
2. The forward (or backward) resistance will change at different range settings due to the effect of the internal series resistance, RS.

Chart 9-2. Test Current vs Range

Range	Max Current	RS*
2000K	0.6 μA	10 MΩ
200K	6 μA	1 MΩ
20K	60 μA	100K
2K	60 μA	100K
200 Ω	600 μA	10K

* In series with the internal 6-V source.

3. Overload conditions are the same as given for resistance measurements.

Measurement of High Resistances—The procedure for measuring resistance greater than 1999kΩ (1.999 MΩ) is given below.

The unknown is connected in parallel with a resistor of known value, 1999kΩ or less, and the resultant is measured.

The unknown is calculated from the relation:

$$Rx = \frac{R_{std} \times R_M}{R_{std} - R_M} k\Omega$$

where,

Rx is the unknown in kΩ
R_{std} is the known resistance in kΩ
R_M is the measured or displayed value in kΩ.

For convenience, Rx can be determined with use of the nomogram shown in Fig. 9-14.

In the example, $R_{std} = 1500$kΩ, $R_M = 1000$kΩ, and Rx = 3000kΩ (3MΩ).

Ballantine 3/24 Digital Multimeter

The Ballantine Model 3/24 digital multimeter (shown in Fig. 9-15) is a full 3-digit instrument. The "full three digits" means the maximum reading is 999, rather than 199 as in a 2½-digit instrument. A top view of the instrument with the case removed is shown in Fig. 9-16. This instrument is battery operated, but an ac supply is available. When batteries are used for power, the expected life of the "throw-away" type of batteries is 300 hours. Power consumption is between 50 and 250 milliwatts (mW). Battery life is extended by using the BRIGHTNESS control, bottom center in Fig. 9-15, to reduce the intensity of the LED readout whenever ambient conditions permit. Automatic polarity is also a feature.

The front panel includes a ZERO screwdriver control (bottom right) to permit exact zero adjustment on all ranges. However, if the ZERO

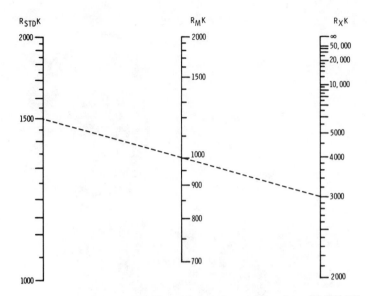

Fig. 9-14. Nomogram for high-resistance measurements (taken from the LDM-850 operating manual).

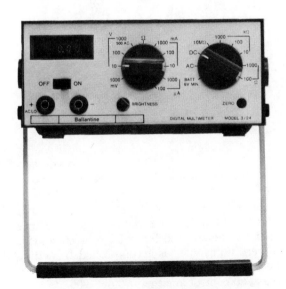

Courtesy Ballantine Laboratories

Fig. 9-15. Ballantine Model 3/24 digital multimeter.

Courtesy Ballantine Laboratories

Fig. 9-16. Top view, case removed, of Model 3/24.

adjustment is used to eliminate the offset of test-lead resistance in the ohmmeter function, the ZERO setting must be returned to normal when the meter is again used for voltage or current measurement. The normal zero adjustment is accomplished by setting the instrument to the 1000-mV range shorting the test leads together, and adjusting the ZERO control to either 000 or −000.

Overloads are indicated by flashing of the most significant digit (which is the left-hand digit). Some optional accessories are an attenuator probe to minimize loading of the test point and to provide a shielded input lead, an rf probe for extending measurement capability up to 500 MHz for sine-wave signals, and a high-voltage dc probe for extending the normal 1000-V dc maximum capability to 30-kV dc maximum.

The following review of the operation of the display, the analog-to-digital conversion, and the signal-conditioner circuits contains information taken from the instrument service manual.

Display—The three display digits, as shown in Fig. 9-15, are monolithic 7-segment LEDs. They are time-shared and sequentially switched at a frequency that exceeds the visual flicker rate. Switching is accomplished by a multivibrator in the power supply. The "on" time of each segment can be varied by adjusting the front-panel BRIGHTNESS control which, in turn, adjusts the duty cycle. Movement of the decimal point is accomplished by setting the front-panel RANGE switch.

Analog-to-Digital Conversion—The adc circuit employs a single negative-going ramp generated by a transistor-capacitor combination. The ramp is reset approximately three times per second—actually every 300 milliseconds—and is applied to two comparator amplifiers. One of these provides a reference, and the other provides the unknown dc analog-voltage-input information. The two comparators feed the MOS LSI which also contains the dc auto-polarity-indicator circuits. The ramp intercepts generate a time interval which is used in the MOS LSI. A clock oscillator provides a frequency which is counted by the LSI during the measurement time interval. One thousand counts accumulate during a full-scale measurement. Display accuracy is maintained through 1200 counts, thus providing 20% overrange.

Signal Conditioners—The signal-conditioner section of Fig. 9-17 includes the input selector switch for voltage, current, and resistance functions, the ac-dc converter, and the dc amp. The ac-to-dc converter is a high-gain differential amplifier which responds to the average value of the incoming ac signal. The instrument actually measures the average value of the input ac but is calibrated in rms volts. The converter is used for both ac voltage and current measurement.

Simpson Model 360 Digital VOM

The Simpson Model 360 digital vom was shown earlier (Fig. 9-1) in this section, and some of its features were described. Here we will examine the instrument, first as an overall system and then as separate sections (the voltage-, resistance-, and current-measurement circuits; the important power-supply features; and the timing diagram).

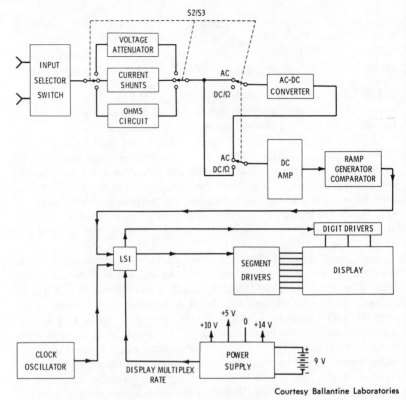

Courtesy Ballantine Laboratories

Fig. 9-17. Block diagram of Ballantine 3/24 digital multimeter.

The Overall System—The simplified basic system diagram for the Simpson 360 digital vom is shown in Fig. 9-18. The signal or quantity to be measured is connected to the input section consisting of the range, function, and signal-conditioning circuits which convert the input into a proportional and numerically equal dc voltage. The dc voltage is applied to the 3½-digit bipolar or dual-polarity dc digital-voltmeter circuit which is indicated on the digital display, and the analog display, and at the analog output. The analog meter provides for convenient and more-rapid indications for nulling, peaking, and varying signal applications—all of which are a little more difficult to interpret when using a digital readout. The power supply provides ac and dc operating voltages. A choice of battery operation is also available by using rechargeable batteries.

DC Voltage Measurements—The dc voltage being measured is applied to the + and COMMON jacks in Fig. 9-19A and is attenuated, if necessary according to the range selected. This voltage is measured by the digital-voltmeter circuit which provides two basic input sensi-

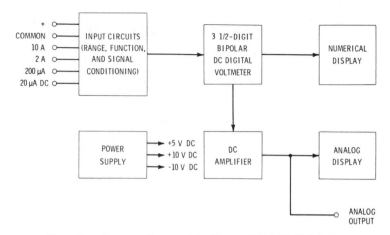

Fig. 9-18. Basic system diagram of the Simpson Model 360 digital vom.

tivities, 200 mV and 2 V, full range; the two input sensitivities make possible a less-complex attenuator. The same attenuator is used on the 20-V and 200-V ranges, and a separate tap is used for the 1000-V range.

AC Voltage Measurements—The basic circuit for measurement of ac voltage is shown in Fig. 9-19B. The ac voltage being measured is applied to the positive and common jacks and is attenuated, if necessary, by the range switch and then applied to the input amplifier. The amplifier output is converted into dc by the ac-to-dc converter and the dc output is applied to the dc digital voltmeter. The same attenuator and dual-sensitivities system used for dc measurement is also used for ac measurement. The converter responds to the average value of the input signal, but its calibration is based on rms sine-wave value. The operational amplifier includes two rectifying diodes and two feedback resistors, R1 and R2, which drive a summing resistor, R3. The diodes and resistors provide negative feedback to the amplifier input. The rectified signal is filtered and the resulting dc voltage is measured by the dc digital voltmeter circuit which, as already mentioned is calibrated to the rms value of the sine wave being measured.

AC and DC Current Measurements—The ac and dc current-measuring circuits are included in Fig. 9-20. The ac current-measuring circuit is essentially the same as the dc current-measuring circuit, except that for ac measurement the voltage developed across the internal shunt resistance is measured by the ac voltage-measuring circuit.

The basic dc current-measuring circuit is shown in Fig. 9-20A. The dc current being measured is connected in series with the

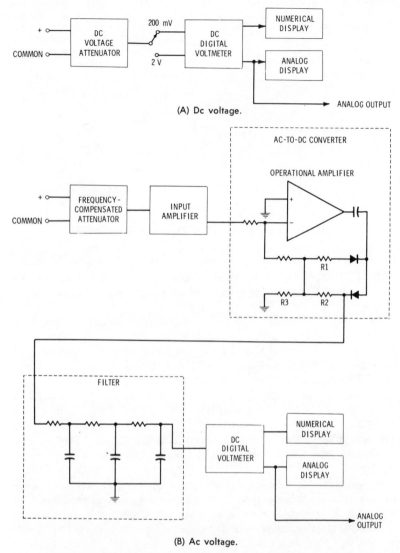

(A) Dc voltage.

(B) Ac voltage.

Fig. 9-19. Basic voltage-measurement circuits of the Simpson Model 360.

common and the appropriate jack or is connected through the range switch and an internal precision shunt resistance. The shunt resistance used depends on the current range selected so that the voltage across the resistance is proportional and numerically equal to the current through it. This voltage is measured by the digital voltmeter circuit which displays a value equal to the current being measured.

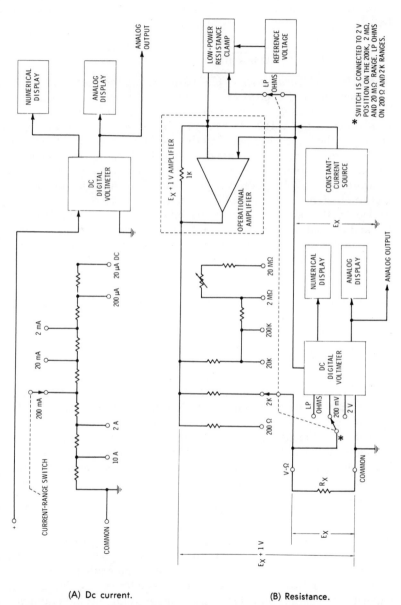

(A) Dc current.

(B) Resistance.

Fig. 9-20. Basic dc current- and resistance-measurement circuits of the Simpson 360.

147

Since the full-range sensitivity of the digital voltmeter circuit is 200 mV, the internal resistance for each current range is equal to 200 mV divided by the full-range current. For example, if the range selected is 200 micramperes, the internal resistance selected would be:

$$\frac{200 \text{ mV}}{200 \ \mu\text{A}} = \frac{200 \times 10^{-3}}{200 \times 10^{-6}} = 1 \times 10^3, \text{ or } 1000 \text{ ohms}$$

Resistance Measurements—The basic circuit for measurement of resistance in the Simpson Model 360 digital vom is shown in Fig. 9-20B. R_x represents the resistance to be measured, and it is connected across the V-Ω and common jacks. A constant current, generated by the meter, is applied to the unknown resistance, the value of the current being determined by the resistance of the range selected. The current causes a voltage drop across the resistor being measured. This voltage drop is proportional to the value of the unknown resistance. The current through unknown resistance R_x is controlled by the operational amplifier which has inputs that "follow" E_x, and the output is always $E_x + 1$ volt. The current from the constant-current source flows through a 1K feedback resistor and provides the 1-volt reference for the amplifier. The ($E_x + 1$ volt) output of the amplifier holds the current through R_x at a constant value, regardless of the value of R_x. The value of the current is determined by which precision resistor, as selected by the ohms range switch, is in series with R_x. The digital voltmeter measures the voltage developed across R_x; the value indicated on the digital readout is equal to the resistance of R_x.

When the digital voltmeter is set for LP OHMS (low-power resistance measurement), the low-power-resistance clamp limits the voltage across the input terminals to 150 mV, open circuit. Thus, the power applied to R_x is limited to less than 100 microwatts. The 200-ohm and 200K settings are the LP OHMS ranges.

Digital Voltmeter Circuit—The digital voltmeter circuit of the Simpson Model 360 vom is shown in a basic block diagram in Fig. 9-21. The dual-slope integration method of analog-to-digital conversion is used. The incoming dc voltage to be measured is sampled and measured at the integrator input at the rate of approximately 5 times per second. The dc voltage to be measured is sampled during an interval, T1, as indicated in the basic timing diagram (Fig. 9-22). The sampling interval is initiated by a pulse from the zero-detector circuit, and the measuring interval, T2, is initiated by an "overflow" pulse from the counter circuits. The timing diagram is based on a positive input signal. During interval T1, integrating capacitor C charges and the output voltage from the integrator increases in proportion with the polarity and magnitude of the input

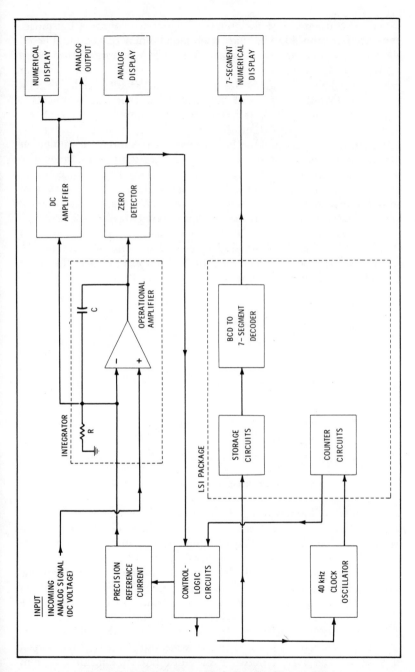

Fig. 9-21. Digital voltmeter block diagram, Simpson 360.

signal. At the same time the counter circuits are counting the pulses generated by the 40-kHz clock oscillator. When the count reaches 8000, the counter circuits generate an overflow pulse and the counters reset to the "zero" state.

The overflow pulse gates the control logic circuit which selects a precision current reference of the required polarity and connects it to the integrator input. At this time, the first integrating interval, T1, ends and the second integrating interval, T2, begins. During T2, capacitor C is discharged by the reference current and the voltage at the output of the integrator decreases toward zero. Simultaneously, the counter circuits are counting the 40-kHz clock pulses, continuing until the integrator output becomes zero. At the crossover of zero, the zero detector generates a pulse which, through the control logic circuits, stops the clock oscillator momentarily and transfers the accumulated count to the storage circuits as an equivalent bcd (binary-coded decimal) signal. The second integrating interval, T2, is now ended; the reference current is disconnected and the cycle repeats, with T1 starting again.

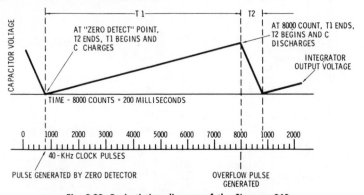

Fig. 9-22. Basic timing diagram of the Simpson 360.

The accumulated bcd signal is converted to signals that the 7-segment decoder/drivers can use for driving the 7-segment numerical display. The value displayed equals the polarity and numerical value of the analog signal at the integrator input.

Power Supply—The block diagram of the power supply for the Simpson Model 360 vom is shown in Fig. 9-23. When the vom is operated from the ac power line, transformer T1 steps down the voltage as required by the rectifier circuit. The rectifier circuit is a full-wave bridge rectifier providing two unregulated dc voltages. One of the voltages is applied to the series regulator with an output of 5 volts. The R1-R2 network provides battery-charging current when the instrument is operated from a power line. The dc-to-dc

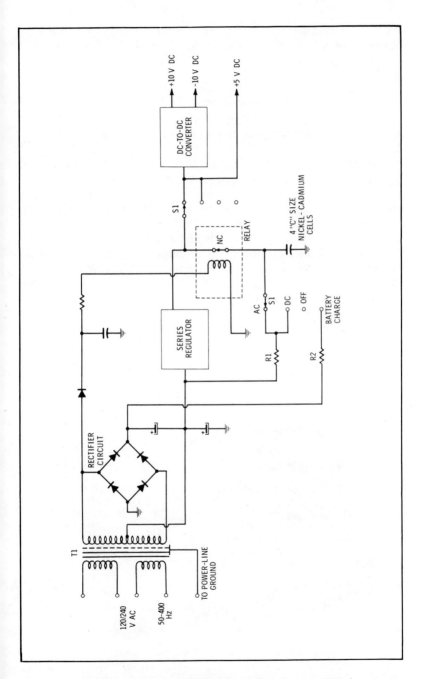

Fig. 9-23. Power-supply block diagram of the Simpson 360.

151

converter changes the 5 volts, whether from a regulated power supply or from a battery, into an ac signal which is stepped up and rectified to provide a regulated output of positive 10 volts and negative 10 volts.

Weston Model 4444 Auto-Ranging Digital Multimeter

The Weston Model 4444 digital multimeter, shown in Fig. 9-24, includes an automatic-ranging feature, a triple-slope adc, and a 4½-digit display.

The theory of operation of the triple-slope conversion process will be explained, according to the manufacturer's service manual, from a functional approach. A complete description of the operating theory, as explained in the manufacturer's instruction manual, is too comprehensive to be included here. The triple-slope converter function diagram is shown in Fig. 9-25, and the waveforms in Fig. 9-26.

The heart of the converter is the integrated circuit, IC, providing all count and control functions. An integrator chain integrates the input voltage and the two reference voltages under control of the IC, and transfers both signal polarity and zero-crossing time to the IC. A time-shared LED display, shown displaying +18888, is driven automatically by the IC, except for a brief blanking period during the drift-correct period for updating the display store. The master

Courtesy Weston Instruments, Inc.

Fig. 9-24. Weston Model 4444 Auto-Ranging digital multimeter.

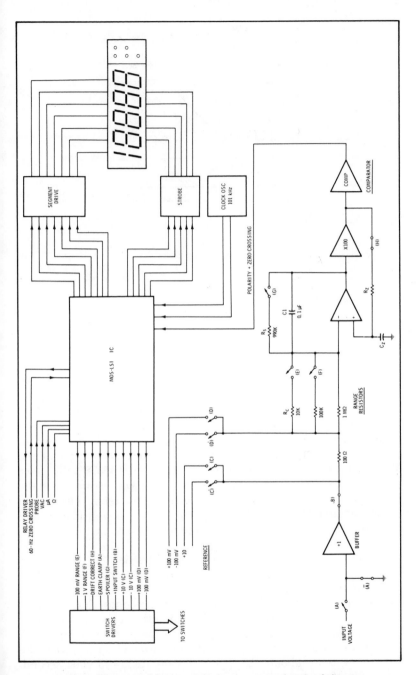

Fig. 9-25. Weston Model 4444 triple-slope converter functional diagram.

153

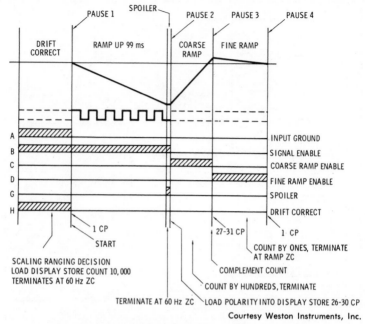

Courtesy Weston Instruments, Inc.

Fig. 9-26. Weston Model 4444 triple-slope converter waveforms.

clock for the system is the 101-kHz crystal-controlled 2-phase oscillator. The switches in the integrator chain are either JFETS or MOSFETS driven by driver circuits which are not shown in Fig. 9-25.

Analog-to-Digital Triple-Conversion Cycle—We will now consider the response of the analog-to-digital triple-conversion cycle of the Weston 4444 to a hypothetical input signal of +3.45678 volts; refer to the converter waveforms of Fig. 9-26. We will assume that the 60-Hz power-line waveform is crossing the zero-reference waveform going negative, but that the conversion cycle is continuous and repetitive. The successive steps are as follows:

1. Pause 1—One clock pulse in duration. This delay is required for proper internal function of the IC. $\overline{A}$ is open; and B is closed, applying zero to the integrator.
2. Ramp Up—10,000 clock pulses (99 milliseconds) long. During this interval A and B are closed, $\overline{A}$ open, applying the input signal and causing the integrator to ramp down. Switches E and F, which control sensitivity, are assumed open so that the ramp rate is about $(3.5 \text{ V}/1 \text{ M}\Omega \times 0.1 \ \mu\text{F}) = 35$ volts/second. The integrator will reach about -3.5 volts in 99 milliseconds.

3. Spoiler—During this interval, switch G closes so that although the input is still applied, the integrator's fall is halted. (The purpose of this step is to improve normal-mode rejection and will be analyzed more fully under Spoiler Operation). Inputs remain connected through switches A and B until the next negative zero crossing of the power line which initiates Step 4.

4. Pause 2—This lasts between 26 and 30 clock pulses, during which the input signal is grounded and the integrator output remains steady. The polarity of the integrator output and hence of the signal is detected by IC during this interval to determine which reference voltages will be subsequently applied, and to store polarity information for subsequent transfer to the display.

5. Coarse Ramp—During "Coarse Ramp," switches A and B are open, and $\overline{A}$ and C are closed applying a −10-volt reference. The integrator ramps back toward zero at 100 volts/second. The counter chain in the IC will have started at a count of zero and will count each clock pulse as 100 counts during this step. Since the voltage is 3.45678 volts, it takes between 3456 and 3457 clock pulses to cross zero, an event which is detected by the comparator and passed onto the IC. The ramp will continue until the first clock pulse after zero crossing. The ramp is then terminated by opening switch C. The count remaining on the IC at this time will be 345700 since the system has been counting by hundreds. The overshoot beyond 345678 is 22 counts, which must be subtracted to produce the desired count.

6. Pause 3—This lasts from 27 to 31 clock pulses and serves to complement the count in the 6-decade register with respect to 999999. The complemented count is 654299. Note that subtraction can be achieved by complementing and adding.

7. Fine Ramp—During this step, switch D is closed, applying a +100-mV reference. The integrator ramps back to zero at 1 volt/second. Since the integrator had overshot zero for 22 microseconds at 100 volts/second, a residue of +2.2 millivolt remains. At the new discharge rate of 1 volt/second, the integrator takes 2.2 milliseconds to reach zero, during which 22 clock pulses are counted. The count is thus brought to 654321 which is internally recomplemented to yield 345678.

8. Pause 4—One clock pulse long—no inputs applied.

9. Drift Correct Period—This final phase has several purposes. First, the final 6-digit stored number is examined and the four most significant figures are transferred to a display store. In this case, the display would read 3456. In the event of a final count 003456, the same four digits would be transferred but decimal-point position and annunciation would be altered to

provide a display of 34.56 mV instead of 3.456 V. (If a number greater than 10999 had been reached, transfer to display would be inhibited and the previous display would continue.)

The six-digit number is also examined to determine if a range change is required. If so, switches E or F or the relay are engaged to change the sensitivity of the system for the next cycle.

During all of Step 9, the input to the integrator is maintained at zero by closing B and A and by opening A. The effect of any offsets that are present due to the unity-gain buffer or that are intrinsic to the integrating operational amplifier are fed back through closed switch H to set up a corrective bias on capacitor Cz. This phase lasts for 10,000 clock pulses in addition to the time taken for scaling and loading the display, thus ensuring adequate time for the corrective bias to be set up on Cz. This bias will maintain itself for the remainder of the succeeding cycle and will be refreshed or updated during the next drift-correct period.

Auto-Ranging—Automatic ranging in the Weston Model 4444 Digital Auto-Ranging Multimeter will be discussed next. This feature makes it unnecessary to select ranges; in fact, only a means for selecting a mode of operation is provided. During the drift-correct period, the 6-digit final count is examined to determine if a range change is needed. The examination takes the form of a search for leading zeros, or overflow beyond 10999. For example, the application of 3.45678 volts when the sensitivity is set at 10 volts ($R_{in} = 1$ megohm) gives rise to a count of 34567 in the 6-digit register. No change is required since the best dynamic range of the integrator is being used. A sudden change to 345.678 millivolts will result in a 6-digit count of 034567 (or 34.5678 millivolts to a count of 003456). This presence of one or two leading zeros causes a shift of one or two places to the left before the zero(s) enters the display store so that only significant data is shown. At the same time, switches F or E, respectively, are closed so that the subsequent conversion will make use of the full dynamic range of the integrator.

Because the interim counts were obtained at an inadequate dynamic range, the last 1 or 2 digits may be in error. However, the reading is transient and recovers to rated accuracy in succeeding conversion cycles. In the interim, a useful display is obtained and the user is not disturbed by the extended inhibit or blanking operation which is characteristic of auto-ranging circuits of earlier design.

If an input larger than 1099 mV but less than 11 volts is applied while the converter is set to maximum sensitivity—i.e., 100 mV ($R_{in} = 10K$—the converter will inhibit one data transfer and will transfer up range to the 10-volt range after which it will display the data. It will either remain at the 10-volt range or go down range once more to the 1-volt range, depending on the actual level. If an input above 11 volts is suddenly applied, the converter will move up range to 10 volts and then up to 1000 volts, after which it will display and either will remain at 1000 volts or will move down range to 100 volts. No more than 2 conversion intervals can be lost due to ranging.

Data-Precision Model 245 Tri-Phasic™ Digital Multimeter

The Data-Precision Model 245 Tri-Phasic™ digital multimeter is a 4½-digit instrument. As shown in Fig. 9-27, it is small in size and can be either battery or power-line operated. According to the manufacturer, it is capable of laboratory accuracy. We will consider briefly its Tri-Phasic™ circuit, which is shown in block-diagram form in Fig. 9-28; its waveform is shown in Fig. 9-29. In the first phase of operation, automatic zero-setting occurs, automatically updating an error integrator/memory circuit, for each conversion cycle. In the second phase, the analog input is integrated together with the stored error, thereby eliminating from the integrated output the offset and drift components that tend to produce errors in certain other digital instruments. The time constant of the integrator is not an accuracy-determining factor. The only accuracy-determining factor is the reference-voltage standard, and the range-divider resistor ratios used only on the higher ranges. In the third phase, the conversion to digital format occurs, returning the integrator to its original zero state, through the original time constant. The Model 245 also includes an Isopolar™ reference source which provides precise +1-V and −1-V reference levels through use of one zener diode especially selected for this function. The zener voltage is

Courtesy Data Precision Corporation

Fig. 9-27. Data Precision Model 245 Tri-Phasic™ digital voltmeter.

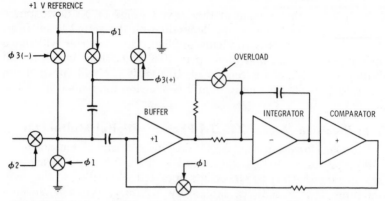

+1 V REFERENCE

OVERLOAD

BUFFER

INTEGRATOR

COMPARATOR

$\phi 3(-)$

$\phi 1$

$\phi 3(+)$

$\phi 2$

$\phi 1$

$\phi 1$

Fig. 9-28. Block diagram of Data Precision Model 245 Tri-Phasic™ digital voltmeter.

either reactively or directly coupled to the analog-to-digital converter in a reversible switching arrangement.

Hewlett-Packard Model 970A Digital Multimeter

The Hewlett-Packard Model 970A digital multimeter, shown in Fig. 9-30 is a portable, hand-held, self-contained digital multimeter providing a 3½-digit readout that can be inverted with the aid of an INVERT DISPLAY switch. The probe tip is built into the end of the instrument. The other test lead is a coiled-pigtail type. A breakdown view of the 970A is shown in Fig. 9-31. Fig. 9-32 shows how convenient it is to use since the instrument can be used in any position and the readout can be inverted, if necessary, by means of the INVERT DISPLAY switch. This instrument features automatic polarity and automatic ranging, ac or dc measurement up to 500 volts, and resistance measurement up to 10 megohms. The test-probe tip is adjustable to several different angles.

The analog-to-digital converter of the Hewlett-Packard 970A includes a dual-slope circuit which is generally the same as the cir-

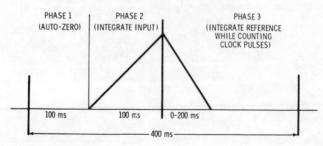

PHASE 1 (AUTO-ZERO)

PHASE 2 (INTEGRATE INPUT)

PHASE 3 (INTEGRATE REFERENCE WHILE COUNTING CLOCK PULSES)

100 ms   100 ms   0-200 ms

400 ms

Fig. 9-29. Waveform for the Tri-Phasic™ circuit of the Data Precision Model 245 digital voltmeter.

cuitry described previously. Referring to the conversion circuit shown in Fig. 9-33 and to the time-interval diagram shown in Fig. 9-34, at time t_1 the unknown input voltage, V_{in}, is applied to the integrator. Capacitor C1 then charges at a rate proportional to V_{in}.

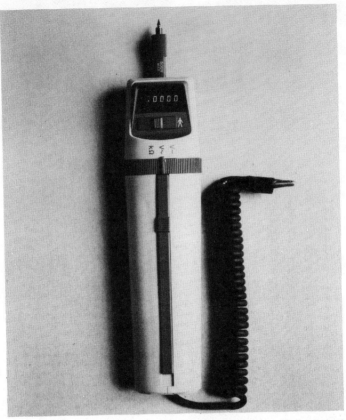

Courtesy Hewlett-Packard

Fig. 9-30. Hewlett-Packard Model 970A.

The counter starts totalizing clock pulses at time t_1, and when a predetermined number of clock pulses has been counted, the control logic switches the integrator input to V_{ref}, a known voltage with a polarity opposite to that of V_{in}. This is at time t_2. Capacitor C1 now discharges at a rate determined by V_{ref}.

The counter is reset at time t_2, and again it counts clock pulses, continuing to do so until the comparator indicates the the integrator output has returned to the starting level, stopping the count. This is at time t_3.

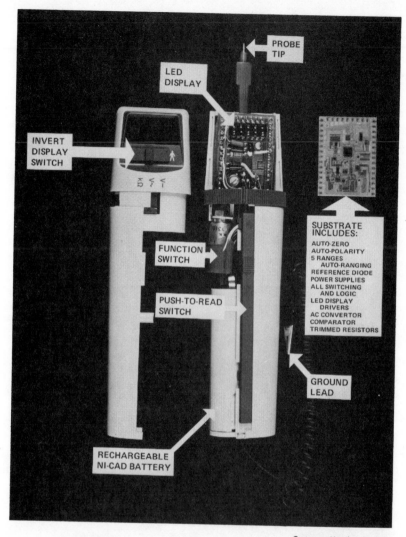

PROBE
TIP

LED
DISPLAY

INVERT
DISPLAY
SWITCH

FUNCTION
SWITCH

PUSH-TO-READ
SWITCH

SUBSTRATE
INCLUDES:

AUTO-ZERO
AUTO-POLARITY
5 RANGES
AUTO-RANGING
REFERENCE DIODE
POWER SUPPLIES
ALL SWITCHING
AND LOGIC
LED DISPLAY
DRIVERS
AC CONVERTOR
COMPARATOR
TRIMMED RESISTORS

GROUND
LEAD

RECHARGEABLE
NI-CAD BATTERY

Courtesy Hewlett-Packard

Fig. 9-31. Breakaway view of the Hewlett-Packard Model 970A.

The count retained in the counter is proportional to the input voltage. This is because the time taken for capacitor C1 to discharge is proportional to the charge acquired, which in turn is proportional to the input voltage. The number in the counter is then displayed to give the measurement reading.

The advantage of this technique over the single-slope technique is that many of the variables are self-cancelling. For example, long-term changes in the clock rate or in the characteristics of the integrator amplifier, resistor, or capacitor affect both the charge and the discharge cycles alike. Considerable long-term deviation from normal values can be tolerated without introducing errors.

Also, since the input voltage is integrated during the up slope, the final charge on C1 is proportional to the average value of the input during the charge cycle. Noise and other disturbances are thus averaged out and have a reduced effect on the measurement. In

Courtesy Hewlett-Packard

Fig. 9-32. Hewlett-Packard Model 970A being used to make measurements on a transmitter.

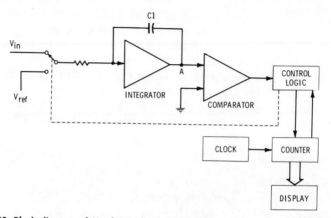

Fig. 9-33. Block diagram of Hewlett-Packard Model 970A analog-to-digital conversion by the dual-slope technique.

particular, by making the charging cycle equal to an integral number of power-line cycles, the effect of any power-line hum is reduced by a substantial amount.

ADVANTAGES AND DISADVANTAGES OF DIGITAL VOMS

Some of the advantages of digital voms have already been discussed or pointed out. A review of some of them is in order at this point. Some of the advantages are: freedom from parallax error, fast and accurate readings, repeatability (different people taking the same reading will come up with the same result), no eye strain, easy to read, no sticky pointer, automatic ranging and overranging, automatic polarity (in most models), and less skill required on the part of the user.

Some of the disadvantages of digital vom's are: difficulty in obtaining the rate or rapidity of change in measured values, less rugged

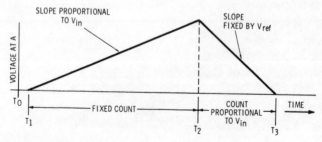

Fig. 9-34. Time-interval diagram for the analog-to-digital conversion in the Hewlett-Packard 970A.

and more easily damaged, usually no facility for decibel measurement, less convenient due to usual larger size and need to be operated from a power line (unless battery operated), difficult to read display under conditions of high surrounding brightness (especially in sunlight), less useful for semiconductor testing (voltage levels on test leads are not high enough to overcome junction areas, so open-circuit readings are often obtained on good devices), more expensive than analog instruments, and response to average values of voltage rather than rms values (analog types respond to rms values).

CARE AND MAINTENANCE OF DIGITAL VOMS

Reasonable care is required in using all digital vom's just as in using all analog voms. However, there are some special directions which should be followed for specific instruments; these special directions can be obtained from manufacturers' service and operating manuals. These usually include directions for removing and replacing batteries and fuses, doing external and internal cleaning, removing the instrument from the case, testing to ensure proper performance, making zero and calibration checks, and making adjustments. Troubleshooting suggestions are also provided in the usual service manual, plus parts lists and information on obtaining parts or repair service from the manufacturer or from a manufacturer's service center. When you are ordering replacement parts or corresponding with the manufacturer about any instrument, you should provide the following information: instrument model number, serial number, part number, schematic symbol number, identification of part, and description of the problem, if appropriate.

QUESTIONS

1. Name four advantages of digital vom's over analog types.
2. Name four disadvantages of digital vom's as compared to analog types.
3. What are the three main stages of a digital vom?
4. Name 4 types of displays used in digital vom's.
5. What is the main advantage of the dual-slope analog-to-digital converter as compared to the single-slope type?
6. In which function of a digital vom is a "constant-current source" employed?
7. How is accuracy usually specified for a digital vom?
8. If a digital vom were designed for 20% overranging on the 1000-volt range, what is the highest value that could be read on this range?
9. Which digit in a digital display is referred to as the most significant?
10. What would be the purpose of a BRIGHTNESS control on a digital vom?

Answers to Questions

CHAPTER 1

1. $0.637 \times 150 = 95.6$ volts, approximately.
2. Zero.
3. $440 \times 0.707 = 311$ volts, approximately.
4. Electrons are negative charges of electricity. They are repelled by other negative charges and are attracted to or toward positive electrical charges.
5. A movement of electrons.
6. 1 ampere equals 1000 milliamperes, 1 milliampere equals 0.001 ampere; 1 ampere equals 1,000,000 microamperes, 1 microampere equals 0.000001 ampere.
7. 50 milliamperes times 0.001 equals 0.050 ampere; 0.00005 ampere times 1,000,000 equals 50 microamperes.
8. Voltage.
9. Battery and electric generator.
10. Electrons move from the negative terminal of the battery, through the electric circuit, to the positive terminal of the battery.
11. Ac voltage also causes electrons to move from negative to positive, but the ac polarity reverses at a regular rate. At first, one of the two ac terminals is negative and the other terminal is positive. After a half cycle of ac voltage has been completed, the first terminal becomes positive and the second one negative. The process repeats at regular intervals. Thus, the ac current moves first in one direction and then in the other.
12. Hertz, abbreviated Hz.
13. 60 Hz.

14. Resistance; it is measured in ohms.

15. Ohm's law expresses the relationship of current, voltage, and resistance in an electrical circuit. The equation for Ohm's law is $E = I \times R$, in which E is voltage in volts, I is current in amperes, and R is resistance in ohms.

16. (a) $I = \dfrac{E}{R}$; (b) $R = \dfrac{E}{I}$.

17. $R = \dfrac{E}{I} = \dfrac{20}{80 \times 0.001} = \dfrac{20}{0.08} = 250$ ohms.

18. $I = \dfrac{E}{R} = \dfrac{100}{25} = 4$ amperes.

19. $E = I \times R = 3 \times 60 = 180$ volts.

20. $P = E \times I = 50 \times 4 = 200$ watts.

21. Peak = rms $\times$ 1.414; $150 \times 1.414 = 212$ volts, approximately.

22. $2 \times 125 = 250$ volts.

CHAPTER 2

1. The d'Arsonval movement.

2. The function switch in the vom is used for setting the instrument to measure ac or dc voltage or current, or resistance.

3. The main parts are a permanent magnet and a coil to which a pointer is attached; the coil rotates in the field of the magnet when current passes through the coil. Also included are the scale plate, or faceplate; a fixed iron core around which the coil is wound; magnet pole pieces; etc.

4. The meter movement should be returned to the manufacturer for replacement if still in warranty, or for repair if out of warranty. First obtain the manufacturer's authorization or direction for returning the movement.

5. The moving coil.

6. Turn the eccentric screw on the outside front of the meter case. This adjustment is provided for in most meter movements—see the instruction manual for the instrument being used.

7. The one requiring 50 microamperes.

8. 20 microamperes (half the full-scale current).

9. The coil probably will be burned out or the meter otherwise damaged.

10. A resistor can be added in parallel with the meter. (A resistor used in this way is called a shunt.)

11. The shunt should have a value 1/9 that of the resistance of the meter movement, or $1/9 \times 500 = 55.5$ ohms. Then 1/10 of 10 mA, or 1 mA, will pass through the meter movement, and 9/10 will pass through the shunt.

12. One or more shunts are provided, and the required shunt is switched into the circuit by means of the range-selector switch.

13. Multiplier resistors are used in series with the meter movement.

14. The total circuit resistance must be $1000 \div 0.001 = 1,000,000$ ohms. The resistance of the multiplier is then $1,000,000 - 1000 = 999,000$ ohms.

15. By including a rectifier in the meter circuit for changing the ac to dc.

16. Any one of the circuits of Fig. 2-9, perferably B or C.

17. Because the resistance of the rectifier and the nature of the rectified current must be taken into account in the ac voltage-measuring function.

18. The pointer will deflect to the half-scale point.

19. Shunt type and series type.

20. Shunt ohmmeter (Fig. 2-10C).

21. For the measurement of output power, generally in audio circuits.

22. Sensitivity depends on the amount of current required for full-scale deflection; less current required indicates higher sensitivity. Sensitivity is specified in terms of *ohms per volt;* a large number of ohms per volt means that a small current is needed to indicate a given voltage, and therefore indicates a sensitive meter.

23. The sensitivity is $\dfrac{1,000,000}{50} = 20,000$ ohms per volt.

24. Usually the sensitivity rating is lower; a vom that has a rating of 20,000 ohms per volt on the dc ranges might have a sensitivity of 10,000 ohms per volt or less on the ac ranges.

25. Loading effect is the change in circuit conditions caused when a measuring device is connected to the circuit. Loading effect of a vom is more noticeable in circuits of high impedance, and in these circuits vom's having higher sensitivity should be used for most accurate measurements.

CHAPTER 3

1. Test-lead wire should be flexible, have good insulation, and should not be susceptible to tangling and snarling.

2. A high-voltage test probe consists of a series multiplier resistor built into the plastic insulating probe. The multiplier resistor permits measurement of voltages higher than those for which a vom is normally designed.

3. The cause of the damage (usually a mistake made in using the vom) should be determined so that the replacement will not be damaged also. An exact-replacement resistor should be obtained from the manufacturer if possible. Most vom's are designed with shunts and multipliers of precise values, of a particular material, or having other characteristics which must be maintained if the vom is to retain its versatility and accuracy.

4. A typical tolerance rating is 1%.

5. R24.

6. Error in reading the scale value caused by being at an angle from the meter.

7. A mirror sometimes is included on the faceplate behind the pointer so that the pointer and its mirror image will coincide when the eye of the observer is directly in front of the pointer.

8. A shunt which some manufacturers make available for use with their particular vom's for measurement of current values in excess of the highest built-in current range of the vom.

9. Test leads should be inspected regularly for good electrical connections at the probes and tips, for loose strands of wire that could cause a shock or a short circuit (or an inaccurate reading), and for breaks in the test-lead insulation.

10. The flange helps to prevent the user from placing his hand too close to the high-voltage point being tested, or too close to the high-voltage multiplier resistor contained near the tip end of the probe.

CHAPTER 4

1. Its loading effect when used for measuring current.

2. This means that when the vom is used for measuring current and when it is deflecting full-scale, it reduces the voltage to the load by 100 millivolts. This reduction is due to the voltage drop across the terminals of the instrument.

3. In low-voltage circuits.

4. Between 2 and 5% for dc ranges, and 2 and 10% for ac ranges.

5. At or near the upper, or full-deflection end of the scale.

6. We could say that resistance measurements are accurate within ±5% of the arc length, or ±5 degrees.

7. The value of the resistance being measured is 150×100, or 15,000 ohms.

8. By noticing the midscale value of the resistance scale and then selecting a range that will give a reading closest to this midscale point for the particular resistor to be measured.

9. When measuring a given amount of voltage, the instrument is accurate within 1 dB at any frequency between 50 Hz and 100 kHz as compared to its reading at 400 Hz.

10. The ability to repeat the same reading for successive measurements of the same quantity.

11. The ability of the vom to indicate accurately at every point on its scale.

12. Most vom's are designed with scales marked in rms ac values, but meters are deflected in proportion to the average value of the ac signal being measured. If the ac voltage is not a sine wave, the reading from the rms scale may not accurately indicate the true rms, or effective, value.

13. Place the instrument in the position in which it is to be used. If the pointer does not rest on zero, adjust the zero-set screw (usually located at the lower center of the front of the meter) for zero indication of the pointer.

14. Plug the red test lead into the jack marked PLUS. Plug the black test lead into the MINUS jack. Set the function switch to dc voltage. Be sure the range switch is set for a range with a full-scale value that exceeds the value of voltage to be measured. Connect the test prods across the two points between which the voltage is to be measured, the red test lead at the more-positive point and the black test lead at the more-negative point.

15. Plug the red test lead into the positive jack, and plug the black test lead into the negative jack. Set the function switch for current. Set the range switch to a current range which will include the value of current to be measured. Be sure the circuit is turned off and any capacitors are discharged. Open the circuit at the point where the current measurement is to be made. Connect the positive test lead to the more-positive circuit point, and the negative test lead to the more-negative circuit test point. Stand clear of the vom. Turn the circuit on and observe the reading on the meter.

CHAPTER 5

1. Use a high-resistance range, for example the R × 10,000 range. The capacitor should be uncharged; if in doubt, short the leads of the capacitor together. Connect the ohmmeter test leads across the capacitor. The pointer should deflect in the direction of zero resistance and then return toward the infinite end of the resistance scale. If the capacitor is shorted, the pointer will go to zero ohms and remain there. A reading near infinity indicates that the leakage resistance of the capacitor is high.

2. Typically about 20 megohms.

3. Electrolytic capacitors have more leakage, or measure lower values of leakage resistance, than do paper or mica capacitors. Also, with the test leads in the correct polarity position, the leakage resistance is much higher than it is with the opposite polarity for the test leads.

4. Use the ohmmeter function of the vom. Connect the test leads across the terminals of the diode and note the resistance reading. Then reverse the test leads and again note the resistance reading. The reading should be high in one direction and low in the other. The high reading might be about 10 times the low reading.

5. With the circuit off and discharged, remove the fuse. Its resistance should be very close to zero ohms if the fuse is good, and infinite ohms if the fuse is open, or bad. If the circuit is live, measure the voltage across the fuse. With a closed live circuit, if the fuse is blown, the voltage across the fuse will be the full voltage of the source. If the fuse is good, the voltage across the fuse will be zero.

6. In a series-string circuit, when one tube develops an open filament, all the tubes in the circuit go out. The full line voltage is then across the terminals of the open filament. The vom can be used as a voltmeter to measure the voltage across each of the tubes in the string. The tube across whose terminals the full voltage is measured is the one with the open filament.

7. The condition of a battery is best tested when the battery is operating under its normal load. Connect the battery to its load (for example, a transistor radio or amplifier). Turn the switch on and measure the voltage across the terminals of the battery. If the voltage measures 75% or less of its rated value, the battery probably should be replaced (or recharged if its the rechargeable type).

8. By using a resistor in series with the "hot" test lead of the vom. The value of the resistance should be selected so that it is high enough to eliminate the "upsetting" effect, but low enough so that a usable reading on the vom can be obtained. The reading with the resistor in series with the test lead will not necessarily be an absolute indication of the actual voltage. (See Table 5-2.)

9. When possible, work with one hand behind you. Be alert to the possibility that faulty operation of the equipment or device being tested can make the device more dangerous than usual to work on. Do not stand on damp or wet surfaces. Keep clear of the circuit. Be sure the power is off when making resistance measurements. When one of the test leads is connected to the live circuit, do not touch the tip of the other test lead.

10. Set the vom function and range switches for measurement of line voltage. Connect one of the test leads to ground. Connect the other test leading to the chassis of the equipment. If a substantial reading is obtained, the chassis is hot.

11. If the pointer does not deflect for any of the functions or ranges, and if the test leads are in good condition, there is a chance that the meter movement is burned out or open. Do not take the meter movement apart; it can be repaired only by the manufacturer. If you know the resistance of the meter movement, and if you are careful, you can set up a test circuit to check the movement as follows.

    Determine the resistance of the meter movement and its rated full-scale current. Connect a 1.5-volt battery and a series resistor to the meter movement. The combined resistance of the series resistor and the meter movement should cause only 2/3 full-scale deflection of the pointer. See Fig. 5-10 and related text.

12. Either the function switch or the rectifier is defective.

13. Only an exact replacement from the manufacturer or other source should be used; otherwise the instrument will not be accurate on the ac ranges.

14. Try not to damage adjacent parts with heat from the hot soldering iron. Beyond this and the use of other ordinary care, special effort should be made not to overheat precision resistors, since this could cause a change in their value.

CHAPTER 6

1. The vtvm generally has a higher input resistance. Also, the vtvm includes an electronic amplifier which increases its sensitivity.

2. The vtvm is less stable and requires warm-up time. It has more-complex circuitry and must be calibrated more frequently. An external source of power is required for the vtvm.

3. (See Fig. 6-5.)

4. The meter movement is "bridged" between the plates of two identical vacuum-tube circuits. When no voltage is being measured, the currents through the tubes are equal and no voltage appears across the meter to cause deflection of the pointer. When the voltages to the grids are not equal, current through the tubes is unbalanced, resulting in current through the meter.

5. It includes a switch and a 1-megohm resistor in series with the test lead. The resistor is switched into the circuit for dc measurements, and is switched out for ac and resistance measurements.

6. It isolates the circuit being checked and reduces the effect of the input capacitance of the vtvm.

7. An rf probe.

8. 1089 megohms is a typical value.

9. By a factor of about 100.

10. The rf probe contains a diode. Rectification of the rf signal takes place in the probe so that the rf signal does not have to travel through the coaxial cable, which would otherwise attenuate the signal.

CHAPTER 7

1. An additional stage of amplification is included.

2. Most recent instruments use selenium or silicon rectifiers; earlier vtvm's used vacuum-tube rectifiers.

3. When the switch is in the "transit" position, a short is placed across the meter terminals. This "damps" (limits or reduces) the movement of the coil and pointer, and prevents damage to the movement and pointer when the instrument is not in use.

4. To prevent accidental changes in the control settings.

5. With the vtvm off, the pointer is set to zero with the zero-set screw usually located on the meter face just below the bottom of the pointer). During operation, the panel ZERO ADJUST control is used in accordance with the operating instructions for the instrument.

6. At least 5 minutes, but 20 to 30 minutes is a better figure. (Consult the manufacturer's instructions for the particular instrument in use.)

7. It is a good practice to install a fresh battery. The output of a weak battery may drop during the course of the calibration procedure and cause an inaccurate calibration.

8. The new tube should be aged for 30 to 50 hours before the vtvm is recalibrated. The instrument can be used during this time, although measurements may not be within rated accuracy.

9. This scale is useful in the alignment of fm-receiver discriminators.

10. The usual reference is 1 milliwatt in 600 ohms. If measurements are made across a different impedance, a correction factor usually must be applied to decibel values read from a scale or chart.

11. The same as the vom: Calibrate it properly before using; replace batteries when they become weak or leaky; observe safety precautions; keep test leads in good condition; store the vom in a safe place, away from machinery, dirty or dusty places, or areas of high temperature or humidity; use exact replacement parts for repair; do not attempt to repair the meter movement yourself—it should be repaired only by the manufacturer.

12. Negligent or accidental misuse of the instrument, failure of component or tube, defect in line cord or ac plug, blown fuse, faulty on-off switch, etc.

13. Recalibrate the vtvm in accordance with the manufacturer's directions.

14. A defective resistor in the multiplier circuit might cause this symptom. (Other possible causes are a weak battery and a weak tube.)

15. By use of an additional external battery and series resistor. (See Fig. 7-9.)

16. The plastic cover over the meter face might have accumulated a static charge as a result of cleaning or polishing of the plastic. The charge can be removed by using a commercially available antistatic solution, or a solution of a good liquid detergent and water. Dampen a clean cloth in the solution and wipe the plastic cover.

CHAPTER 8

1. They are small, lightweight, compact, battery operated, portable, versatile, and they require no warm up.

2. Field-effect transistor.

3. High input impedance.

4. Basically, the functions of the gate, source, and drain correspond to those of the grid, cathode, and plate, respectively, of a triode vacuum tube.

5. See Fig. 8-2.

6. For checking the condition of the batteries that power the instrument.

7. 10 to 15 megohms.

8. It provides a high input impedance so that the circuit under test is not loaded.

9. Roughly six months to a year. Battery life can be extended by turning the instrument off when it is not used and by not storing it in a warm location.

10. From 0° F downward, battery capacity decreases, but batteries become completely inoperative at −20°F.

CHAPTER 9

1. No parallax error; fast, accurate readings; repeatability; no eye strain; easier to read; no sticky pointer; automatic ranging and over-ranging on some models; automatic polarity on some models; and less skill required.

2. Changing values harder to interpret, less rugged, does not usually make decibel measurements, less convenient due to size and usual operation from power line, harder to read in high surrounding brightness, limited usefulness for semiconductor testing, and greater cost.

3. Signal conditioner, analog-to-digital converter, display.

4. LED, gas discharge, incandescent, and Nixie.

5. Less susceptible to noise, etc.

6. Resistance-measuring function.

7. Percentage of reading plus or minus one digit.

8. 1200 volts.

9. The one farthest to the left.

10. To control the brightness of the display.

Index

1863